Cybersecurity Fundamentals: Building Blocks For A Secure Future

Disclaimer:

The information provided in this book is for general informational purposes only. While every effort has been made to ensure the accuracy and completeness of the information, the publisher and author assume no responsibility for errors, omissions, or damages resulting from the use of the information contained herein. The reader is advised to consult with a professional for any specific cybersecurity concerns or requirements.

Trademarks:

All trademarks mentioned within this book are the property of their respective owners. The use of trademarks is solely for identification and does not imply endorsement, association, or sponsorship.

Note:

Please be aware that this book is intended to provide general guidance and information on cybersecurity fundamentals. It is not a substitute for professional advice or a comprehensive guide to specific cybersecurity practices. The reader should exercise their own judgment and seek the advice of professionals when implementing cybersecurity measures. The publisher and author disclaim any liability arising directly or indirectly from the use of this book.

Chapter 1. Introduction To Cybersecurity

Welcome to the interconnected realm of the 21st century, a dimension where data has become the lifeblood, and our lives are entwined with an invisible network of bits and bytes. Stepping into this realm reveals an essential protective shield: cybersecurity.

Imagine a bustling metropolis. The infrastructural sophistication, the hum of energy, the life pulsating through every vein of the city. Now, perceive cybersecurity as the guardian standing tall at the city gates, armed with the tools to shield the metropolis from the threats of the chaotic outside world.

Cybersecurity, at its most fundamental, is about preserving the integrity, confidentiality, and availability of information. It forms the backbone of modern enterprises, safeguarding critical data against an ever-evolving landscape of threats. Like a chameleon, cybersecurity adjusts and adapts to the changing contours of the digital world, developing new defences as new threats emerge.

With a meteoric rise in our dependence on digital technology, the fortress of cybersecurity has grown from a luxury to a necessity. From the personal data stored on your smartphone to the confidential information held by multinational corporations, cybersecurity's influence permeates every facet of our digital lives.

Yet, the concept of cybersecurity remains enigmatic for many, often seen as a labyrinth of complex jargon and technicalities. This perception, while not unfounded, serves as a barrier that can deter engagement and understanding. However, as we delve deeper into this book, the fog will gradually clear.

So, welcome to your journey through the vast, intricate world of cybersecurity—a labyrinth, yes, but one rich with intrigue, challenges, and the promise of a safer digital future.

1.1. What is Cybersecurity?

Cybersecurity, a term frequently tossed around in conversations about the digital world, is not merely a concept—it's a multidimensional, intricate system, teeming with complexities. Yet, at its core, cybersecurity embodies a simple goal: protection. Protection of our digital lives, our data, our privacy, and, ultimately, our society.

At its most basic definition, cybersecurity is a set of practices, processes, and technologies designed to safeguard systems, networks, and data from digital attacks. These threats, often termed as 'cyber threats', seek to exploit vulnerabilities in the system to illicitly access, alter, or delete sensitive data, disrupt operations, or even extort money.

However, the labyrinth of cybersecurity is not confined merely to these technical definitions. Its true essence, its real understanding, comes from the understanding of the three key pillars upon which it stands— confidentiality, integrity, and availability, often abbreviated as the CIA triad.

Confidentiality involves ensuring that sensitive information is accessed only by authorised individuals. This protection may be achieved through numerous ways, such as data encryption, two-factor authentication, or strict access controls.

Integrity relates to maintaining and assuring the consistency and accuracy of data. It ensures that the information is reliable and accurate throughout its entire lifecycle, and that no unauthorised person has tampered with it.

Various tools like checksums, hashes, and digital signatures help uphold data integrity.

Availability is about ensuring reliable access to information and resources to authorised individuals when needed. Through redundancy, failover, and backup practices, availability ensures the smooth and efficient operation of systems and networks.

These three pillars form the bedrock of cybersecurity, supporting every strategy, practice, and solution in this realm. However, understanding them is only scratching the surface of what cybersecurity truly entails. The domain's real depth unfolds as we begin to navigate the various types of cybersecurity—network security, application security, information security, operational security, disaster recovery, and business continuity planning, to name a few. Each type represents a unique facet of the cybersecurity gem, working together to form a robust and effective Defence mechanism.

The landscape of cybersecurity is a dynamic one, always in flux, always evolving. The emergence of new technologies like artificial intelligence, machine learning, and quantum computing continually redefine its boundaries. At the same time, they also introduce new vulnerabilities, creating a perpetual arms race between cyber-defenders and cyber-attackers.

The world of cybersecurity is also strongly interwoven with legal and ethical considerations. Laws around data protection, privacy, cybercrimes, and more influence the practice of cybersecurity. The need for ethical hacking—a practice where cybersecurity professionals intentionally penetrate networks and systems to identify weaknesses—highlights the importance of ethical considerations in this space.

Grasping the idea of cybersecurity is not about memorising definitions or understanding complex algorithms. It is about understanding its essence—protection in the digital realm. The next time you log in to your email, perform an online transaction, or simply browse the web, remember the guardian that shields you. That guardian is cybersecurity, the protector of the digital metropolis.

1.1.1. Definition and Key Elements

A paradigm of modern technology, cybersecurity, encapsulates the vast ecosystem of defending electronic systems, networks, and data from malicious digital assaults. Broadly defined, it signifies the techniques and methodologies employed to protect the integrity, confidentiality, and availability of data.

Diving deeper into this technological sphere, we find key elements that underline cybersecurity's ethos and methods. These pillars of cybersecurity, the 'CIA' triad, represent the cardinal principles guiding the strategies and protocols within this digital Defence discipline.

The first of these elements, Confidentiality, is a commitment to preventing unauthorised access. It ensures that privileged information remains concealed from those without the appropriate permissions. Encryption, access controls, and network traffic analysis constitute a few methods to preserve confidentiality.

Next in line, Integrity, guarantees that the data remains unaltered and trustworthy throughout its lifecycle. This pillar ensures that any change, whether accidental or deliberate, is detected and ideally prevented. Hashing algorithms, digital signatures, and version control tools are commonly used techniques to sustain data integrity.

The third pillar, Availability, ensures that systems, data, and resources are accessible to authorised users when needed. From maintaining hardware, performing regular updates and patches, to creating reliable backups, the efforts aimed at maintaining availability are crucial for smooth operations.

Beyond the CIA triad, additional elements play a significant role in shaping cybersecurity's landscape. These include:

Authentication, verifying the identity of users, systems, and devices. It is typically achieved through passwords, biometrics, and digital certificates.

Authorisation, deciding what an authenticated entity is allowed to do. Role-based access control is a popular authorisation strategy, delineating system access permissions based on roles within an organisation.

Accountability, ensuring actions within a system can be traced back to an entity, creating a trail that discourages malicious activity and aids in the investigation of incidents.

Non-repudiation, providing proof of communication or data transfer, preventing entities from denying their actions. Digital signatures are a common tool to enforce non-repudiation.

Finally, addressing cybersecurity is incomplete without mentioning its continuously evolving nature. The emergence of cloud computing, Internet of Things (IoT), and AI has led to an increasingly interconnected digital landscape, expanding the horizons for cybersecurity. It's a perpetual journey, keeping pace with emerging technologies and ever-evolving threats, crafting newer, stronger shields for our digital lives. In this digital era, cybersecurity is not a luxury—it's a necessity.

1.1.2. Goals of Cybersecurity

Like any discipline, cybersecurity is guided by a set of overarching goals that inform its strategies and actions. These goals act as navigational markers, aligning the efforts of cybersecurity professionals towards securing digital ecosystems.

At the heart of cybersecurity's objectives lies the protection of information and systems from damage and disruption. This ranges from safeguarding individual privacy to ensuring the continuity of critical infrastructure. The loss or compromise of data can lead to significant financial loss, reputational damage, or even physical harm in the case of critical systems, such as those used in healthcare or transportation.

Further, cybersecurity strives to enable safe operation in the cyberspace and ensure data integrity, confidentiality, and availability. It seeks to maintain a digital environment where individuals, organisations, and governments can operate with the assurance that their sensitive data and systems are secure. The goal is not merely to respond to threats but to proactively design systems and protocols that can withstand attacks.

Additionally, one of cybersecurity's primary goals is to prevent unauthorised access. It aims to keep malicious actors at bay, barring them from accessing information or systems without appropriate permissions. This objective underscores the importance of robust access control mechanisms in cybersecurity.

Cybersecurity also aims for rapid and effective response and recovery after an incident. Despite the best preventive measures, breaches can occur. When they do, the objective is to detect them swiftly, minimise damage, isolate the threat, and restore normal operations as quickly as possible. Incident response and disaster recovery plans are integral to achieving this goal.

Cybersecurity's goals also extend to compliance with laws and regulations. Organisations must comply with a variety of cybersecurity regulations, from data protection laws to industry-specific standards. Compliance not only minimises legal risks but also often aligns with best practices for data protection.

The overarching aim is to create a safe and trustworthy digital environment. Cybersecurity goals are not static. As technology and threats evolve, so too will the objectives of cybersecurity, adapting to new challenges and striving to stay a step ahead of potential threats.

1.2. The Importance of Cybersecurity

Cybersecurity isn't just a buzzword; it's a societal necessity. As technology permeates every aspect of our lives, the importance of cybersecurity escalates. It's not merely about preventing attacks and data breaches but also about preserving trust and enabling progress.

The cornerstone of the digital age is data. From personal information to intellectual property, from banking transactions to healthcare records, data fuels the digital ecosystem. However, this valuable resource is a prime target for cyber threats. Cybersecurity ensures the protection of this data, preserving its confidentiality, integrity, and availability.

Another critical facet of cybersecurity is its role in ensuring business continuity. Cyber-attacks can cripple organisations, disrupting operations and causing significant financial losses. By safeguarding systems and networks, cybersecurity allows organisations to operate with minimal disruption and maintain their reputation among stakeholders.

On a larger scale, cybersecurity safeguards critical infrastructure like power grids, healthcare systems, and financial institutions, many of which are prime targets for cyber threats. Disruptions in these sectors can lead to severe societal consequences, further highlighting the importance of cybersecurity.

In the context of individual privacy, cybersecurity is fundamental. With vast amounts of personal data being stored and processed digitally, protecting this information is paramount. Cybersecurity tools and practices protect individuals from identity theft and other forms of cybercrime, empowering people to use digital services safely.

Cybersecurity also plays a vital role in enabling digital innovation. As organisations embrace emerging technologies such as AI, IoT, and cloud computing, they expose themselves to new cyber threats. Cybersecurity evolves alongside these technologies, providing the necessary safeguards that allow for secure innovation.

Lastly, cybersecurity is crucial in upholding legal compliance. Many jurisdictions have enacted data protection laws, and non-compliance can lead to hefty penalties. Cybersecurity ensures that organisations adhere to these regulatory requirements, mitigating legal risks.

In an increasingly connected world, the significance of cybersecurity cannot be overstressed. It serves as the bedrock of digital trust, empowering individuals, businesses, and societies to harness the power of technology securely and confidently.

1.2.1. Cybersecurity for Individuals

In the digital landscape where individuals are increasingly living, learning, working, and playing, cybersecurity takes on a personal dimension. The security practices adopted by individuals can spell the difference between secure digital experiences and devastating cyber incidents.

At its core, cybersecurity for individuals revolves around protecting personal information. Social security numbers, bank account details, health records, and even personal photographs—these are all lucrative targets for cybercriminals. Identity theft is a common consequence of such breaches, leading to financial losses, damaged credit, and lengthy recovery processes.

Cybersecurity measures help individuals maintain control over their personal data and avoid these outcomes.

Beyond protecting personal data, cybersecurity for individuals also involves securing digital devices, from laptops and smartphones to smart home devices. Cyber threats can transform these devices into tools of espionage or disruption. For instance, malware can transform a laptop into a botnet node used for launching attacks, while unsecured IoT devices can be exploited to gain unauthorised access to networks or personal information.

Online behaviours also fall within the purview of personal cybersecurity. Phishing attacks, for instance, rely on individuals making mistakes—clicking on a malicious link, for example. Cybersecurity education helps individuals recognise these threats, fostering safer online behaviours and minimising the chances of successful cyber-attacks.

Moreover, cybersecurity for individuals means safeguarding personal privacy. From surveillance by intrusive advertising to more sinister forms of stalking, strong cybersecurity practices can help individuals maintain their privacy in the face of these threats.

In essence, cybersecurity for individuals is about personal empowerment. It's about equipping individuals with the tools, knowledge, and practices needed to navigate the digital world securely, protecting their data, devices, and digital experiences from cyber threats. It's about ensuring that the promises of the digital age—connectivity, convenience, and opportunity—are not undermined by cyber risks.

1.2.1.1. Personal Data Protection

Personal data is a treasure trove, a goldmine brimming with nuggets of information that, if misused, can bring about a nightmare of intrusion and identity theft. As such, protecting this invaluable asset becomes paramount in maintaining an individual's cybersecurity.

Personal data protection, in essence, is a facet of cybersecurity that focuses on safeguarding an individual's private information. It stretches across various forms of data - financial details, health records, social security numbers, email addresses, phone numbers, and more.

Cybercriminals salivate at the thought of these digital riches, using a range of strategies, from spear-phishing attacks to brute force hacking, to pilfer this information. Such breaches can lead to a series of personal crises: identity theft, unauthorised transactions, and even blackmail.

In shielding oneself from these potential disasters, a blend of proactive and reactive strategies is key. On the proactive side, robust passwords, multi-factor authentication, and the use of encrypted networks help build a fortress around one's personal data. At the same time, it is important to monitor financial accounts and credit reports regularly for signs of suspicious activity.

Moreover, it's crucial to understand the perils of oversharing online, which often creates easy pickings for cybercriminals. By learning to discern what information should remain private and managing digital footprints, individuals can significantly reduce the risk of their data falling into the wrong hands.

Finally, being aware of the rights provided by data protection regulations helps individuals exercise control over their personal data. It empowers them to make informed decisions about which organisations to trust with their information and how to respond if a data breach occurs.

Ultimately, personal data protection is about establishing and maintaining a secure zone around one's most sensitive information, thereby ensuring a safer digital journey.

1.2.1.2. Protecting Digital Assets

The vault of digital assets held by individuals has grown exponentially, broadening the landscape of what needs protection in the realm of cybersecurity. From a meticulously curated music library to valuable cryptocurrency investments, from heart-warming family photos to that essential work project, our digital lives are awash with assets worth safeguarding.

Protecting digital assets is a task that spans beyond mere ownership. It requires an understanding of the inherent value these assets hold, and the risks associated with their digital nature. A mishap can lead to the loss of something irreplaceable, or a cyber-attack might strip away an asset's value overnight.

A comprehensive strategy for digital asset protection begins with secure storage. Backing up assets on external hard drives or using cloud storage services with a strong reputation for security, is a wise first step. It's essential to ensure that these backups occur regularly and automatically, if possible, to minimise the risk of recent data loss.

Next is data encryption, which converts digital assets into unreadable text, decipherable only with the correct key. By encrypting files, individuals can ensure that even if an unscrupulous party obtains them, they remain as indecipherable as an ancient, lost language.

Strong passwords serve as the first line of Defence, yet they are often overlooked. Crafting unique and complex passwords for different accounts minimises the risk of unauthorised access. Employing a reliable password manager to manage these can reduce the burden of remembering them all.

In the case of financial assets like cryptocurrencies, the use of cold wallets - storage that isn't connected to the internet - can drastically reduce the threat of cyber theft.

Remember, knowledge is power. Staying informed about the latest cyber threats and the best protective measures is key to securing digital assets.

1.2.2. Cybersecurity for Organisations

As the world grapples with an increasing reliance on digital technologies, organisations - from small businesses to multinational corporations - have become targets of sophisticated cyber-attacks. Each successful breach not only jeopardises sensitive data but also risks damaging an organisation's reputation, consumer trust, and financial stability.

Organisations handle a vast array of data, from confidential customer details to proprietary intellectual property. In the hands of cybercriminals, such information can lead to financial fraud, identity theft, or competitive disadvantage. The sheer magnitude and value of the data make cybersecurity an essential component of organisational strategy.

A successful cybersecurity framework within an organisation goes beyond technology; it involves people and processes. Each employee, from the CEO to the newest intern, has a role in maintaining cybersecurity. Awareness and training programs are crucial to create a culture of cybersecurity and to counter phishing attempts or other user-targeted attacks.

Processes should be in place to identify potential threats, respond to security incidents, and recover from breaches. Incident response teams are a vital cog in this machinery, trained to mitigate damage and quickly restore normal operations.

Moreover, as organisations increasingly rely on third-party services, supply chain cybersecurity has gained prominence. Ensuring the security of all partners is as important as fortifying the organisation's own Defences.

Technologically, organisations should deploy robust firewalls, regularly update software to patch vulnerabilities, use intrusion detection systems, and conduct frequent security audits. The use of artificial intelligence and machine learning can help predict and counter threats more efficiently.

Finally, regulatory compliance is integral to organisational cybersecurity. Adhering to data protection regulations not only ensures the organisation's legality but also signals its commitment to cybersecurity to its stakeholders.

As our digital footprint expands, the need for robust cybersecurity measures in organisations continues to escalate. An effective cybersecurity framework is no longer a luxury; it's a necessity, a core pillar supporting the integrity of our digital infrastructure.

1.2.2.1. Business Continuity

The digital world, while bustling with opportunities, is also fraught with risks that can disrupt business operations. Cybersecurity plays a critical role in ensuring business continuity, allowing organisations to maintain their services and meet their obligations amidst potential disruptions.

When a cyber-attack breaches an organisation's Defences, the repercussions can be far-reaching. Ransomware can freeze operations, data breaches can expose confidential information, and a Distributed Denial of Service (DDoS) attack can overload systems, bringing them to a standstill. Cybersecurity measures aim to prevent such incidents or minimise their impact when they occur.

Strong cybersecurity protocols facilitate swift recovery from these incidents. Using practices such as data backup, disaster recovery planning, and incident response strategies, organisations can resume operations with minimal downtime. For instance, having a recent backup allows quick restoration of systems following a ransomware attack.

Moreover, cybersecurity plays a significant role in preserving the reputation of an organisation. Consumers, partners, and shareholders value a business that demonstrates resilience against cyber threats. Maintaining trust in the face of adversity contributes to business continuity.

Business continuity planning that includes cybersecurity measures also helps organisations meet regulatory obligations. Regulations such as the General Data Protection Regulation (GDPR) in the EU mandate certain levels of data security and incident response. Compliance shows due diligence and aids in avoiding costly penalties.

The interplay between cybersecurity and business continuity is a dance of resilience. By protecting data, systems, and operations, cybersecurity practices help ensure that businesses can withstand cyber threats and continue to thrive in the ever-evolving digital landscape.

1.2.2.2. Protecting Customer Data and Trust

Customer data becomes a vital resource and a considerable responsibility. Safeguarding it isn't merely an organisational duty but a requirement to uphold the trust customers place in businesses. Cybersecurity finds its core relevance in this act of safekeeping.

Imagine every piece of customer data as a precious gem in a digital treasure trove. From personal information to financial details, these gems attract the attention of cybercriminals. Their goal is to pilfer these valuables, turning the breach into a lucrative venture while leaving organisations and customers in disarray.

Cybersecurity is the fortress protecting this treasure. It deters potential intruders, detects their nefarious activities, and deflects their assaults. It ensures that customer data remains confidential, untampered, and accessible only to those authorised to view it.

The trust relationship between customers and businesses is crucial. Customers share their data, believing that companies will protect it. A breach in cybersecurity is not just a failure of technology—it signifies a broken trust. Therefore, cybersecurity measures play an indispensable role in fortifying this relationship, promoting customer confidence and loyalty.

Furthermore, many regulatory bodies worldwide mandate stringent data protection rules. Compliance to these standards is vital to avoid severe penalties and maintain organisational reputation. Thus, cybersecurity is not only an ethical requirement but also a legal one, underscoring the gravity of protecting customer data and trust.

The labyrinth of the digital world can seem daunting, but robust cybersecurity measures light the path. They protect the valuable trust customers place in businesses, ensuring that this sacred bond remains unbroken. Through vigilant protection of customer data, organisations can confidently navigate the digital landscape, nurturing enduring customer relationships.

1.2.2.3. Regulatory Compliance

Just as lighthouses guide ships through tumultuous seas, so too does regulatory compliance serve as a beacon in the complex ocean of cybersecurity. It denotes the adherence to rules and laws that govern the way organisations handle data and maintain the security of their information systems. A necessary companion in the journey of cybersecurity, regulatory compliance is not just about fulfilling statutory obligations but about steering the ship safely and responsibly.

In an age marked by an incessant rise in cybercrimes, governments and international bodies across the globe have tightened the reins. They've laid down meticulous guidelines to safeguard sensitive information. Rules like the General Data Protection Regulation (GDPR) in the European Union, California's Consumer Privacy Act (CCPA), and others across different jurisdictions have taken centre stage. These regulations establish the responsibilities of organisations in protecting data and lay down stringent penalties for non-compliance.

Adhering to these rules isn't just about dodging fines or penalties—it's about upholding an organisation's integrity. Regulatory compliance helps to imbibe best practices in cybersecurity, ultimately fortifying the organisation's Defence mechanism. It ensures a systematic approach to data protection, establishing a shield that mitigates risks of cyber threats and breaches.

Moreover, demonstrating compliance bolsters an organisation's credibility. It signifies to stakeholders—be it customers, partners, or investors—that the organisation respects and values data privacy and security. It shows that the organisation navigates the digital sea not just with technological prowess but with ethical grounding and responsibility.

Thus, regulatory compliance stands at the intersection of law, ethics, and cybersecurity. It's a lighthouse that guides organisations in their cyber voyage—illuminating the path, forewarning dangers, and ultimately leading them towards a safe harbour. And cybersecurity is the ship, fortified and made resilient by following the guiding light of compliance. Together, they ensure a journey that respects boundaries, upholds responsibilities, and stays course in the vast ocean of digital data.

1.3. The Evolution of Cybersecurity

If one were to step back in time to the dawn of digital computing, the notion of cybersecurity would appear rather alien. The concept, as we understand it today, was not part of the original blueprint of the internet. It emerged only as the digital realm evolved and expanded, forever altering our lives in ways that once seemed the stuff of pure science fiction.

The earliest inklings of cybersecurity can be traced back to the 1970s when computer systems began to network, giving birth to the potential for 'unauthorised access'. This term, so commonly used today, was a nascent concept back then. The years that followed were marked by an escalating game of cat and mouse between emerging hackers and the nascent security protocols designed to stop them.

As computers infiltrated homes in the 1980s, the playground of potential targets grew exponentially. Cybersecurity began to take on a more recognisable shape, as software companies created the first antivirus programs to protect individual users from malicious attacks.

The following decade, the 1990s, witnessed the rapid expansion of the internet, taking cybersecurity into uncharted territories. Companies raced to connect, opening up myriad vulnerabilities along the way. As e-commerce blossomed, the importance of securing transaction data became paramount, leading to the development of encryption technologies and the SSL (Secure Sockets Layer) protocol.

Moving into the new millennium, the landscape transformed dramatically with the advent of social media, cloud computing, and mobile devices. These developments exponentially increased the amount of data produced, shared, and stored online. Cybersecurity now had to protect not just organisations or individuals, but an entire interconnected ecosystem of users and devices.

Fast forward to the present day, and cybersecurity has become a multi-faceted discipline, grappling with a staggering array of threats. From state-sponsored cyberattacks and espionage to ransomware attacks and data breaches, the challenges are many and varied. AI and machine learning now play pivotal roles in both, launching sophisticated cyberattacks and defending against them.

At each stage of this evolution, the cybersecurity landscape has mirrored the broader digital realm: from protecting single computers, to networked systems, to cloud platforms, to the current focus on securing interconnected digital ecosystems. It's a journey that has seen cybersecurity morph from a niche concern to a societal necessity.

It's a story still being written, with many twists and turns yet to come. Emerging technologies like quantum computing, the Internet of Things (IoT), and decentralised networks are the next frontier. The evolution continues, and as we stand on the brink of these new technological leaps, one thing remains clear: cybersecurity, as the guardian of our digital world, will continue to be an indispensable part of the narrative.

1.3.1. From Antivirus Software to Holistic Cyber Defence

Retracing the steps of cybersecurity, one is struck by the journey from basic antivirus software to the comprehensive, holistic Defences in play today. Once upon a time, cybersecurity was but a thread in the vast tapestry of technology, one primarily aimed at shielding individual computers from malware.

Antivirus software—conceived during the 1980s—epitomises this early phase of cybersecurity. It functioned as a kind of digital armour, designed to guard against specific threats by identifying and nullifying harmful programs. This reactive approach was effective for the simpler, less aggressive digital threats of that era.

Yet, as the digital landscape morphed and expanded, so too did the threats. The age of the internet saw cyber threats grow in complexity and sophistication, requiring a more strategic approach. Firewalls emerged as the next line of Defence, controlling traffic to and from a network, and intrusion detection systems began monitoring for any unusual activity that could indicate a breach.

While these security measures marked progress, they still operated under a distinct disadvantage—they were inherently reactive. They waited for an attack to occur before springing into action. The turn of the century, however, witnessed a paradigm shift: from Defence to deterrence. Proactive cybersecurity measures, such as penetration testing and vulnerability assessment, came into the picture, seeking to identify and fix security gaps before they could be exploited.

Today, cybersecurity has evolved into a more holistic and proactive discipline, integrating elements of risk management, regulatory compliance, incident response, and user education. Contemporary strategies stress the importance of a layered Defence model, often visualised as an onion, with multiple overlapping layers of security measures.

The centrepiece of this modern approach is a robust security culture built on education and awareness, one that seeks to weave security into the very fabric of an organisation's operations. In this sense, every employee becomes a part of the organisation's Defence mechanism, equipped with the knowledge and tools to identify, and prevent potential cyber threats.

Yet, the evolution is far from over. Cybersecurity is moving beyond the confines of organisations, striving for a more unified, collaborative approach. This concept, often termed as Collective Defence, envisions a world where organisations, irrespective of their size or industry, actively share threat intelligence and collaborate on Defence strategies.

Thus, the story of cybersecurity's evolution is a tale of constant adaptation, a relentless quest to stay one step ahead of an ever-changing threat landscape. From humble beginnings in antivirus software, it has grown into a multi-dimensional discipline that continues to evolve alongside our digital world, striving for a more secure tomorrow.

1.3.1.1. The Era of Antivirus Software

The advent of antivirus software marked the inception of digital Defence. This era, flourishing throughout the 1980s and 1990s, was a time when the cyberspace was relatively uncharted, where the concept of a 'digital attack' was still a novel and unfamiliar notion.

The underlying principle of these early Defences was reactive prevention. Antivirus software would be armed with a database of known viruses and threats. Once detected, these harmful entities would be isolated and eliminated. The early antivirus systems were like sentinels standing watch, armed with the knowledge of previously identified foes, ever prepared to defend their digital territory.

However, the digital world of this era was notably less complex. The digital threats mainly constituted worms and viruses that often did little more than cause a nuisance. Most were not designed to steal data or spy on users but rather to highlight the vulnerabilities within systems. The creators of these nuisances were often tech enthusiasts testing boundaries rather than criminals with malicious intent.

Though antivirus software provided a substantial line of Defence, it was, in essence, a catch-up game. Every new threat required an update to the antivirus databases, and many systems were compromised before these updates could be implemented. In retrospect, it's clear that this was a rudimentary first step, the beginning of a much longer and more complex journey in the ongoing endeavour to protect digital spaces and assets.

1.3.1.2. Modern Cyber Defence Strategies

The progression from the antivirus software era to the present day represents a paradigm shift in cybersecurity approaches. From a game of catch-up, the focus has pivoted towards anticipating attacks and building resilience. Today, the scope of cybersecurity extends far beyond just defending against known threats. It now incorporates robust strategies and cutting-edge technology to proactively protect, detect, respond, and recover from a myriad of digital attacks.

The modern cybersecurity framework takes a more holistic approach. Threat intelligence is a crucial component of this framework, providing real-time insights about emerging threats. Unlike the era of antivirus software, where threats were known, modern cybersecurity must contend with Zero-day attacks – attacks that exploit unknown vulnerabilities.

Moreover, the advent of machine learning and artificial intelligence has dramatically transformed cybersecurity strategies. Predictive analytics capabilities can identify patterns and anticipate potential threats before they materialise, marking a significant advance from the reactive methods of yesteryears.

Another critical strategy of modern defence involves the concept of 'defence in depth'. Instead of a single line of defence, there are multiple layers to protect the system. Should an attack bypass one layer, the next is ready to halt its progress.

Security audits and penetration testing have also become integral components of the cybersecurity framework, offering a proactive method of identifying and addressing vulnerabilities. By simulating cyber-attacks, organisations can identify weak points and take remedial actions before actual attacks occur.

Finally, education and awareness have been given an increased emphasis. In the modern digital world, every user is a potential weak link. Hence, educating all users about potential threats and safe practices is key to building a more secure digital landscape.

In essence, modern cyber defence strategies represent a new generation of digital shields, with an emphasis on anticipation, layered defence, ongoing assessment, and user education. These proactive strategies have evolved to protect digital spaces amidst an increasingly sophisticated and rapidly changing threat landscape.

1.3.2. The Increasing Complexity of Cyber Attacks

Cyber threats have moved in tandem with the digital world's evolution, growing in sophistication and complexity. The challenges posed by this new breed of cyber-attacks are more significant, straining the conventional security measures and pushing for continuous innovation in cybersecurity strategies.

In the early days, cyber threats largely involved relatively simple malicious software, mostly aimed at causing inconvenience. Today, the landscape is dramatically different. Attackers are often highly skilled, innovative, and backed by substantial resources. They are capable of orchestrating intricate attacks designed not only to exploit known system vulnerabilities but also to exploit previously unknown flaws, or "zero-day" vulnerabilities.

Multi-stage and polymorphic malware represent some of the advancements in cyber threats. In a multi-stage attack, the hacker typically infiltrates the target in phases, bypassing defences and establishing footholds to further the assault. Polymorphic malware, on the other hand, can alter its code to evade signature-based detection systems, making it especially challenging to detect.

Distributed Denial-of-Service (DDoS) attacks have also seen increased complexity. Today's DDoS attacks often involve multiple vectors, targeting various parts of the infrastructure simultaneously to maximise disruption.

Perhaps even more concerning is the rise of Advanced Persistent Threats (APTs). These sophisticated, long-term attacks are typically aimed at exfiltrating data rather than causing immediate damage. The subtle nature of these threats makes them especially hard to detect and counter.

State-sponsored cyber-attacks also represent a critical evolution in cyber threats. Often these attacks are well-funded, sophisticated, and driven by political, economic, or military motivations.

Lastly, cybercriminals increasingly use social engineering methods to manipulate individuals into compromising their own security. Phishing, baiting, and pretexting are among the popular techniques used, capitalising on human errors to breach defences.

In a world where the complexity of cyber-attacks continues to escalate, it becomes critical to stay one step ahead. The constantly evolving labyrinth of digital threats underscores the need for organisations to remain vigilant, continuously updating and strengthening their cybersecurity protocols and systems.

1.3.2.1. Evolving Tactics of Cybercriminals

Cybercriminals have always thrived on adaptability, with their methods constantly shifting in response to advancements in digital defence mechanisms. This section uncovers the evolution of their strategies, painting a picture of the changing face of digital predators.

One of the fundamental shifts in tactics has been the migration from mass, opportunistic attacks to targeted, individualised schemes. Earlier, cybercriminals often cast a wide net, spreading viruses, or launching attacks on any susceptible system. This has gradually changed, with modern attackers often choosing specific targets, conducting meticulous planning, and crafting individualised attack vectors.

The introduction of social engineering represents a significant change in cybercrime tactics. As technical defences become more robust, cybercriminals increasingly turn to manipulating human behaviour. By masquerading as a trustworthy entity, they trick users into revealing sensitive information or performing actions that compromise security. This trend illustrates the shift from a purely technical battleground to a psychological one.

Moreover, the use of ransomware has surged in recent years, posing an enormous threat to both individuals and organisations. This malicious software encrypts a victim's data and demands payment for its release. Cybercriminals have found it to be a profitable venture, given the vital importance of data and the often-limited defences of targets.

Cybercriminals have also adopted advanced techniques such as exploiting zero-day vulnerabilities and leveraging fileless malware. Zero-day exploits take advantage of software flaws unknown to the software's creator, making them particularly difficult to defend against. Fileless malware resides in a system's memory rather than on the hard drive, leaving no file to be scanned and detected.

More recently, cybercriminals have begun harnessing the power of Artificial Intelligence (AI) and Machine Learning (ML) to automate their attacks. They can analyse patterns, find vulnerabilities faster, and customise their attacks more effectively. This burgeoning trend presents an entirely new level of cyber threat.

These evolving tactics of cybercriminals signify the dynamic nature of cybersecurity. As threats become more sophisticated, a combination of evolving technology, policy, and user education becomes vital to maintaining an effective defence. The continued study and understanding of these ever-changing strategies are paramount to staying a step ahead in the cybersecurity game.

1.3.2.2. State-Sponsored Cyber Attacks

Cybersecurity, once a primarily technical concern, has become an issue of national security. The rise of state-sponsored cyber-attacks underscores this shift, revealing a new frontier in geopolitical conflict - the digital realm.

State-sponsored cyber-attacks are actions conducted or sponsored by a nation-state aimed at disrupting the digital infrastructure of another nation, corporation, or group of individuals. These activities can serve a variety of purposes, from stealing classified information and disrupting critical infrastructure to sowing discord and influencing political outcomes.

The impact of these cyber-attacks extends beyond the virtual world. In many cases, they can have tangible real-world consequences, particularly when they target critical infrastructure such as power grids, transportation systems, or healthcare facilities. Such attacks can lead to physical damage, disruptions in essential services, and potentially loss of life.

One notable aspect of state-sponsored cyber-attacks is their complexity and sophistication. They often involve advanced persistent threats (APTs) that infiltrate target systems, remaining undetected for extended periods while continuously monitoring activities and exfiltrating data. These threats are typically highly sophisticated and employ a variety of techniques to avoid detection and maintain persistence.

The rise of state-sponsored cyber-attacks also brings into focus the issue of attribution. Pinpointing the origin of a cyber-attack can be exceedingly difficult due to the use of proxies, botnets, and other methods to conceal the attacker's identity and location. This has led to the development of new methods for cyber forensics and attribution, often involving the correlation of technical evidence with geopolitical analysis.

Lastly, this new landscape of digital warfare has led to a growing dialogue about the need for international norms and treaties regarding state behaviour in cyberspace. Just as traditional warfare is governed by international law, there is a growing consensus that similar rules should apply to state actions in the digital realm.

The rise of state-sponsored cyber-attacks serves as a stark reminder that the landscape of conflict is continuously evolving. Cybersecurity has become inextricably linked with national security, and in the face of this reality, nations must prioritise the strengthening of their cyber defences and work towards establishing international norms for behaviour in cyberspace.

1.4. Key Cybersecurity Terminology

1.4.1. Common Terms and Acronyms

Understanding cybersecurity requires familiarising oneself with the terms that define its landscape. Here is a brief glossary of essential cybersecurity terms:

Malware: This term, short for "malicious software," refers to any program or file that is harmful to a computer user. Types of malware include viruses, worms, Trojans, and spyware.

Phishing: A type of cyber-attack where the attacker poses as a reputable entity to deceive individuals into providing sensitive information such as usernames, passwords, and credit card numbers.

Firewall: A network security system that monitors and controls incoming and outgoing network traffic based on predetermined security rules. Firewalls act as a barrier between a trusted and an untrusted network.

Encryption: The process of converting information or data into a code to prevent unauthorised access. It's the key method for securing data in transit and at rest.

VPN (Virtual Private Network): A service that lets you access the web safely and privately by routing your connection through a server and hiding your online actions.

DDoS Attack (Distributed Denial of Service): An assault on a network that floods it with so much traffic that it becomes unavailable to users.

Ransomware: A type of malware that encrypts a victim's files. The attacker then demands a ransom from the victim to restore access to the data upon payment.

Two-Factor Authentication (2FA): An additional layer of security used to ensure that users are who they claim to be. It requires not only a password and username but also something that only the user has on them (like a piece of information only they should know or have immediately on hand - like a physical token).

Zero-day vulnerability: A software security flaw that is known to the software vendor but doesn't have a patch in place to fix the vulnerability. Attackers exploit zero-day vulnerabilities through zero-day attacks.

Intrusion Detection System (IDS): A system that monitors networks for suspicious activity and issues alerts when such activity is discovered. It's a vital component of a robust cybersecurity strategy.

Understanding this terminology lays a crucial foundation for deepening one's knowledge of cybersecurity, as they define the threats, defences, and strategic considerations that characterise this vital field.

1.4.2. Understanding Threat Actors and Their Motivations

In the chess game of cybersecurity, it's important to not only understand the mechanics of the game, but also to know the players on the other side. A "threat actor" refers to any individual or group that attempts to harm a computer system, often for personal or financial gain, political interests, or even for the thrill of disruption. Here are the common types of threat actors and their usual motivations:

Cybercriminals: This group is primarily motivated by financial gain, employing tactics like ransomware, identity theft, phishing, and credit card fraud. They range from lone wolves to organised crime groups, and their targets can be anyone from individuals to large organisations and financial institutions.

Hacktivists: As the portmanteau suggests, these actors combine hacking with activism. They often attack systems to advance political or social objectives, using methods like DoS attacks to disrupt services and data breaches to expose information.

Insiders: These actors can be current or former employees, contractors, or business associates who have legitimate access to an organisation's systems. Their motivations can vary wildly, from financial gain, revenge, thrill-seeking, to even inadvertent actions without harmful intent.

Nation-states: These are typically advanced threat actors backed by a national government, aiming to perform espionage, sabotage, or disruption. Their motivations are generally political, military, or economic, and their targets are often other nations' critical infrastructures, defence systems, or key industries.

Terrorists: Cyber-terrorists use cyber-attacks to cause widespread disruption and fear, often for ideological or political reasons. They target critical infrastructure like power grids, transportation systems, or communication networks.

Advanced Persistent Threats (APTs): These actors, often nation-states or criminal organisations, aim to infiltrate network infrastructure to spy on, steal data from, or disrupt their targets. They are characterised by their patience, sophistication, and long-term approach.

Understanding threat actors and their motivations isn't about generating fear; rather, it's about gaining clarity. It allows individuals and organisations to better anticipate potential threats, to empathise with the human aspects of cyber conflict, and to formulate more effective defence strategies.

CHAPTER 2. CYBER THREATS AND ATTACKS

Our journey into the cyber realm now brings us face-to-face with the complex beasts that lurk in its shadows: cyber threats and attacks. This chapter pulls back the curtain on the chilling tactics employed by threat actors as they hunt for vulnerabilities and exploit them for their gain. The aim here isn't to induce a state of paranoia but rather to instil a sense of awareness and preparedness, arming you with the knowledge to better understand, anticipate, and mitigate these cyber threats.

We delve into the realm of malware, from ransomware to spyware, exploring their characteristics and consequences. We'll also traverse the terrain of social engineering attacks, revealing how human psychology can become the weakest link in cybersecurity. We will further navigate through the landscape of network attacks, explaining their methodologies and how they can affect infrastructure.

Cyber threats and attacks continually evolve, mirroring the rapid pace of technological development. As we dissect their anatomy, it's essential to remember that each section of this chapter is a piece in the broader jigsaw puzzle of cybersecurity. Understanding these threats and how they operate is the first step towards building an effective cyber defence.

2.1. Understanding the Threat Landscape

Navigating the labyrinth of the internet is like journeying through an exotic digital jungle. The dense foliage of codes and data conceals numerous threats lurking in the shadows. These threats evolve with time, mutating and growing stronger as technology progresses, creating a rapidly shifting threat landscape.

In essence, the cyber threat landscape encompasses the myriad of potential hazards present in our interconnected digital world. These threats range from common malicious software (malware) to sophisticated nation-state sponsored cyber espionage. This landscape is a dynamic space, influenced by technological advances, evolving user behaviour, and regulatory changes.

Malware, one of the most prevalent threats in this landscape, is software specifically designed to disrupt, damage, or gain unauthorised access to a computer system. They come in many forms, each with unique characteristics and damaging capabilities. Viruses, worms, and Trojans infect systems, disrupt operations, and sometimes lay dormant until triggered. Ransomware, a subset of malware, encrypts user data and demands ransom in exchange for the decryption key.

Another prominent feature of this landscape is social engineering attacks. Cybercriminals often use manipulative tactics to trick users into revealing sensitive information or performing actions that compromise security. Phishing attacks, a common type of social engineering, deceive users into clicking on malicious links or attachments, often disguised as legitimate emails or messages.

In more sophisticated corners of this landscape, we find targeted attacks such as Advanced Persistent Threats (APTs). These threats are typically orchestrated by well-resourced groups (like nation-states) and are aimed at stealing information or disrupting operations over a prolonged period.

The threats mentioned above thrive on vulnerabilities. Vulnerabilities refer to weaknesses in a system that can be exploited. They may exist in hardware, software, or even in the behaviour of users. By exploiting these vulnerabilities, threats can bypass security measures, access sensitive information, and disrupt systems.

For instance, software vulnerabilities, often caused by coding errors or design flaws, can allow unauthorised access to a system if successfully exploited. Similarly, hardware vulnerabilities can lead to physical damage or unauthorised access to data stored on hardware devices. User behaviour vulnerabilities, such as weak passwords or uninformed clicking on suspicious links, remain a significant factor in successful cyber-attacks.

However, the threat landscape is not just about the threats themselves but also the strategies and tactics they use. Cyber threats are not static; they continually evolve to bypass security measures and exploit new vulnerabilities. For example, malware has seen significant evolution, from simple viruses that caused minor disruptions to ransomware capable of crippling entire networks.

As we delve deeper into this chapter, we'll dissect the specifics of these threats, delve into their mechanisms, and explore the real-world impacts they've had. The aim is to provide a detailed understanding of the hazards we face in the digital jungle and equip you with the knowledge needed to navigate it safely. By appreciating the breadth and depth of this landscape, we can design and implement security measures that effectively guard against these threats.

2.1.1. The Concept of Attack Surfaces

Visualise a bustling coastal city. In terms of cybersecurity, the city's coastline is the 'attack surface,' the shoreline where a potential invader might make landfall. It's the sum total of points where the city can be breached or attacked. The broader the coastline, the higher the potential vulnerabilities and consequently, the higher the risk of an attack.

In the realm of cybersecurity, the concept of attack surfaces takes a similar analogy. It represents the myriad of points in a system where an unauthorised user could potentially infiltrate or extract data. The broader the attack surface, the more the opportunities for a cyber-attacker to exploit vulnerabilities, penetrate defences, and access valuable data.

Attack surfaces are not limited to merely technological aspects; they extend to people and processes as well. For instance, an attacker could exploit a software vulnerability, use social engineering to trick an employee into revealing sensitive information, or take advantage of an outdated process to bypass security measures.

There are three primary types of attack surfaces:

Digital attack surface: This is the most common and includes software vulnerabilities in an organisation's digital infrastructure. Examples include insecure code, unpatched software, misconfigured servers, or vulnerabilities in web and mobile applications.

Physical attack surface: This includes all the physical points of access to an organisation's network and its devices. An attacker could gain physical access to servers, network equipment, or workstations, or use wireless attacks to infiltrate the network.

Human attack surface: People are often the weakest link in cybersecurity. This attack surface includes any vulnerability arising from human behaviour. Examples include falling victim to a phishing attack, using weak passwords, or unintentionally downloading and executing malicious software.

Mitigating threats to these attack surfaces demands a combination of technology, people, and processes. On the technology front, practices like patching and updating software, robust code review processes, and strong network security can reduce the digital attack surface. For the physical attack surface, measures such as secure workstations, surveillance systems, and access control mechanisms are vital. As for the human attack surface, the solution lies in continuous cybersecurity education and awareness training to promote safer online behaviour.

The concept of attack surfaces is crucial in understanding the threat landscape because it helps identify the areas at risk and where defences should be fortified. As the technological realm expands with the integration of more devices, cloud services, and advanced technology, the attack surface also broadly expands, calling for ever-evolving defence mechanisms.

In the sections to come, we will break down each of these attack surfaces, the vulnerabilities they contain, and the strategies to mitigate these vulnerabilities. By comprehending the attack surface and its components, we move a step closer to creating a secure cyber environment.

2.1.2. Identifying Vulnerabilities and Exploits

Picture yourself in an intricate maze of cybersecurity, where each turn might present an opportunity or a pitfall. On one side, there are vulnerabilities—weak spots or gaps in your cyber defences. On the other side, there are exploits—tactical manoeuvres that adversaries use to seize these vulnerabilities.

Vulnerabilities are the chinks in the cyber armour, the spots in the software or system where the defences are not quite robust. They can arise from a variety of sources—outdated software, misconfigured servers, software bugs, or even oversights in design and implementation. Essentially, any fault that leaves the door ajar for an attacker to breach your defences can be classified as a vulnerability.

Exploits, on the other hand, are like the master keys crafted by cyber adversaries to unlock these vulnerabilities. An exploit can be a piece of code, a chunk of data, or a sequence of commands that take advantage of a vulnerability to cause unintended behaviour in a system—typically leading to data breach, data corruption, or a takeover of the system.

Identifying vulnerabilities is like solving the first half of the puzzle in cybersecurity. It involves proactive measures such as vulnerability scanning, penetration testing, code review, and security audits. Vulnerability scanning employs automated software to inspect a system against known vulnerability signatures. Penetration testing takes this a step further by attempting to exploit the vulnerabilities to determine exposure. Code review and security audits are more thorough inspections to ensure robust software and processes.

But identifying vulnerabilities is only half the battle won. Once vulnerabilities are identified, they must be promptly remediated to prevent exploitation. Patch management plays a crucial role in this process— updating systems with patches released by software vendors to fix the identified weaknesses. This game of cat and mouse between discovering vulnerabilities and patching them is ongoing, often referred to as the 'vulnerability management lifecycle.'

However, not all vulnerabilities can be patched, and not all can be discovered before an attacker does. This is where the concept of 'zero-day' vulnerabilities comes in, which are vulnerabilities unknown to the parties responsible for patching or fixing them. In such scenarios, the exploits leveraging these vulnerabilities are known as 'zero-day exploits.'

As we tread further into the maze of cybersecurity, we must always remember: understanding and navigating vulnerabilities and exploits are not mere optional routes. They form the essential pathways leading to a more secure cyber environment. The journey may seem complex, but each vulnerability identified and remediated, every exploit understood and defended against, brings us one step closer to the exit of the maze, into a realm of enhanced cybersecurity.

2.2. Common Types of Cyber Threats

As we peel back the layers of the digital world, we find ourselves at the heart of an ever-evolving battleground—a landscape fraught with threats that continuously challenge the sanctity of our cybersecurity fortresses. This section presents a panoramic view of this digital arena, highlighting some of the common types of cyber threats that lurk in its shadows.

Just like soldiers in a real-world battle, cyber threats come in different forms, each armed with unique strategies and tactics. Their objectives may range from petty cybercrime to grand-scale espionage. But no matter how big or small, these threats pose significant risks, disrupting the smooth operation of digital systems and compromising data integrity and confidentiality.

As we delve deeper into this realm, we will explore a variety of common threats, from those as prevalent as phishing and malware to more sophisticated ones like advanced persistent threats (APTs) and ransomware. Each of these cyber threats possesses distinct characteristics and impacts and understanding them is the first step toward building an effective defence strategy.

Navigating through this digital battleground is no less than stepping into the unknown, with threats emerging from the most unexpected quarters. But with the right knowledge and tools at our disposal, we can turn this uncertainty into a manageable challenge. Welcome to the diverse landscape of cyber threats—a journey that will arm us with a nuanced understanding of our adversaries and equip us with the necessary shield and sword to safeguard our digital territories.

2.2.1. Malware: Viruses, Worms, Trojans, Ransomware

Peering into the world of malware, we encounter a host of unwelcome digital parasites - viruses, worms, trojans, and ransomware. Each is a malicious software, carefully engineered to infiltrate, damage, and disrupt digital environments. They are the hooded figures that lurk in the shadows, waiting for the perfect opportunity to strike.

A computer virus, like its biological counterpart, infects healthy systems and replicates itself. It spreads by attaching itself to clean files and, in turn, corrupts them. The corrupted files then become carriers, leading to a widespread infestation if left unchecked.

Worms, though similar to viruses, have a sinister edge. They can replicate and propagate without user intervention, swiftly consuming system resources and causing severe network disruptions. Unlike viruses, they don't need to attach to a file to spread; they travel unaided, wreaking havoc along their path.

Trojans, on the other hand, are the masters of deception. Named after the legendary Greek wooden horse, they masquerade as legitimate software, fooling users into willingly installing them. But once inside, they create backdoors, allowing cybercriminals unrestricted access to the infected system.

Last, but certainly not least, we confront ransomware, the hijackers of the digital world. Upon infection, ransomware encrypts files, effectively locking users out of their own systems. The attackers then demand a ransom for the decryption key—a scenario akin to a hostage situation.

The varying nature of these malicious software underpins their danger. Each employs distinct strategies to breach defences, exploit vulnerabilities, and disrupt operations. Yet, despite their differences, they share a common goal: to cause damage and chaos in the digital realm. Understanding the mechanics of these unwelcome guests, their modes of operation, and the threats they pose, is an integral part of formulating an effective cybersecurity strategy.

2.2.1.1. *Understanding Malware Mechanisms*

Piercing the veil of malware's deceptive tactics illuminates the cunning strategies and crafty mechanisms that underpin their successful propagation and impact. This enlightenment is not just for the technically adept; it is vital knowledge for all digital inhabitants.

Malware is engineered to be elusive. It often operates covertly, maintaining a low profile while it embeds itself and performs its insidious tasks. The signature attributes include self-replication and propagation, system alteration, information collection, and evasion of detection. Each type of malware uses a unique blend of these traits to navigate and conquer digital environments.

Let's examine the common mechanisms more closely.

Self-replication and propagation are key characteristics of many malware types, particularly viruses and worms. They can create copies of themselves and transfer between systems, often exploiting network connections or software vulnerabilities.

System alteration typically involves changing settings, modifying data, or damaging files. This action often impacts the system's operation, potentially crippling essential functions or creating vulnerabilities for further exploitation. Trojans, for instance, often create backdoors—unauthorised access points—in systems they infect.

Information collection, a familiar trait of spyware, involves harvesting data from the infected system. This data can be anything from login credentials to personal information, and it often serves as a valuable commodity in the shadowy realms of the cyber black market.

Evasion of detection is arguably the most critical malware characteristic. Malware frequently uses sophisticated methods to hide from antivirus software and system security mechanisms. Advanced malware can even alter its code or behaviour to evade signature-based detection systems.

Understanding these mechanisms aids in appreciating the ingenuity behind malware and the threats they pose. It also emphasises the need for robust and dynamic cybersecurity strategies, capable of countering these ever-evolving tactics.

2.2.1.2. Malware Detection and Prevention

Malware detection and prevention are like an endless chess match between cybercriminals and cybersecurity professionals, each side seeking the upper hand in this complex and ever-evolving game. This strategic battle takes place on a board that expands with the increase of digital assets and attack surfaces, as each new software, device, or system added represents another square on the board that needs protection.

Two primary approaches for detecting malware exist: signature-based detection and behaviour-based detection. Signature-based detection is akin to recognising the face of a known criminal; it works by matching malware to known patterns of malicious code or "signatures" in a database. While highly effective against known threats, its limitation is that it may fail to detect novel malware or variations of existing ones, created specifically to evade these signatures.

On the other hand, behaviour-based detection observes the actions of software within a system. If certain actions correlate with known malicious behaviour, the software is flagged as potentially harmful. This approach can detect unknown malware or variations by focusing not on how they look (their code), but on what they do.

Prevention involves a multitude of strategies, both technical and behavioural. Technically, the consistent updating and patching of systems and software are essential. This is because updates and patches often contain fixes for vulnerabilities that malware may exploit. Adept use of firewalls and the careful configuration of network permissions can also limit malware propagation.

The human element cannot be overlooked either. Cyber hygiene practices, such as being cautious of unknown emails or links, using strong, unique passwords, and being wary of suspicious online behaviour, go a long way in preventing malware infections.

In this perpetual chess match, understanding detection methods and preventive strategies helps individuals and organisations make informed moves, fortifying their defences, and keeping malicious opponents in check. However, the game never ends. It's an evolving cycle requiring constant vigilance and proactive measures to stay ahead of potential threats. As technology advances, so do the sophistication and cunning of malware. Thus, cybersecurity professionals must be nimble, ready to innovate and adapt as they strive to maintain the balance in this digital contest of strategy and tactics.

2.2.2. Phishing and Social Engineering Attacks

Phishing and social engineering attacks are the modern-day equivalents of old school con artists' tricks but updated for the digital age. Their weapons of choice are not lockpicks or sleight of hand, but a detailed understanding of human psychology, an ability to exploit trust, and the audacity to masquerade behind the anonymity of the internet.

Phishing is a method in which attackers disguise as trustworthy entities, usually through emails or text messages, luring their victims into revealing sensitive data. These data often include usernames, passwords, credit card numbers, or other valuable information. A well-crafted phishing email can be a masterclass in deception, mimicking not just the look and feel of a legitimate communication from a reputable source, but also the tone and language that inspire trust.

On the broader stage, social engineering attacks encompass phishing and go beyond, exploiting human behaviours and social interactions. Preying on natural tendencies such as trust, curiosity, or fear, these attacks might involve impersonating an authority figure, manipulating a person into bypassing security measures, or even creating a fake crisis to create urgency and prevent clear thinking.

The damage inflicted by these attacks can be severe. Stolen credentials can lead to identity theft, financial loss, and breach of personal or business data. Also, the emotional toll on the victims, who are often deceived into betraying their own privacy or their organisation's security, should not be underestimated.

Protecting against phishing and social engineering attacks is challenging because it's essentially a human problem, not a technical one. While spam filters and anti-phishing software can provide some level of defence, education is the most potent weapon. By promoting awareness of these tactics, fostering a culture of scepticism towards unsolicited communication, and maintaining strict protocols for the verification of identities and the handling of sensitive information, individuals and organisations can construct a robust defence against these deceptive practices.

The battle against phishing and social engineering is more than just a technical challenge—it's a psychological one. By understanding the devious art of these attacks, we can better prepare ourselves and stay one step ahead of the digital con artists.

2.2.2.1. Techniques Used in Phishing Attacks

Phishing attacks come in various disguises, each a calculated manoeuvre designed to manipulate, deceive, and extract valuable information. Here, we delve into some commonly used tactics in the phishing attacker's playbook.

At the heart of every phishing attack is the art of impersonation. Attackers often mask their identity by pretending to be a trustworthy entity - it could be a reputable company, a government institution, or even a familiar individual. They meticulously design their emails or messages to mimic the legitimate source, from the color schemes and logos to the email addresses and signatures.

One of the common techniques is link manipulation. Here, the attacker crafts a link that appears safe but redirects the user to a malicious website. Often, the actual URL is obfuscated by subtle misspellings, long strings of meaningless characters, or cleverly designed subdomains that may not raise immediate suspicion. These malicious sites usually mimic the look of legitimate ones to trick users into entering their credentials or other sensitive information.

Another widely used technique is website forgery. In such cases, attackers create a counterfeit version of a legitimate website, tricking users into believing they are interacting with the real deal. These websites often exploit browser vulnerabilities to install malware or steal information.

Phishing attacks can also involve social manipulation. Scare tactics are frequently used, with attackers sending alarming messages about account compromises or fraudulent activities, prompting immediate action. By creating a sense of urgency, they hope to override the victim's caution and push them into hasty decisions.

Spear-phishing and whaling represent more targeted attacks. In spear-phishing, attackers focus on a specific individual or organisation, tailoring their phishing attempts with personalised information to enhance credibility. Whaling goes a step further, targeting high-profile individuals within an organisation, like CEOs or CFOs, with potential for larger pay-outs.

Finally, attackers use baiting techniques, promising enticing rewards to trick users into divulging their personal information. This could involve offering free services, winning an unexpected lottery, or other too-good-to-be-true propositions.

2.2.2.2. Defending Against Social Engineering

The rise in social engineering attacks underlines a critical truth - the human element can often be the weakest link in the cybersecurity chain. Therefore, enhancing this link is crucial. Here's how we can reinforce our defences against social engineering.

Awareness is our first line of defence. Regular training programs and awareness campaigns can help individuals understand the tactics used by social engineers, the tricks they employ, and the disguises they use. It's essential to stay updated about the latest scams, the trending lures, and the new forms of attacks that emerge.

In the face of a potential attack, exercising caution is key. For instance, before clicking a link or opening an attachment, verify its authenticity. Pay attention to subtle red flags, like spelling mistakes, unfamiliar email addresses, or requests for sensitive information. If you're unsure, it's always better to refrain from responding and consult an IT or security professional.

Two-factor authentication (2FA) is a valuable tool in our defence arsenal. By requiring a second form of identification, it adds an extra layer of protection. Even if an attacker manages to obtain your password, the chances of them also having access to your second form of verification, like a physical token or biometric information, are significantly lower.

Implementing good internet habits is another way to fortify our defences. This includes creating strong, unique passwords for each account, regularly updating and patching systems and software, and ensuring that personal and financial information is shared only over secure and encrypted connections.

Organisations can also make use of technological solutions to detect and block social engineering attacks. These can include spam filters, phishing detection systems, and secure email gateways.

Lastly, developing a culture of cybersecurity within the organisation is crucial. Encourage open communication about potential threats. If someone spots a suspicious email, they should feel comfortable reporting it. A collective, informed vigilance is one of the most effective ways to thwart social engineering attempts.

Through these steps, we can strengthen our human firewall, reducing the effectiveness of social engineering attacks and safeguarding our valuable information.

2.2.3. DDoS Attacks

Imagine an avalanche, a torrent of snow barrelling down a mountain, engulfing everything in its path, making the landscape unrecognisable. DDoS attacks function in a similar way, overwhelming systems with traffic until they become inaccessible, swept away in a digital avalanche.

DDoS attacks are initiated by multiple sources, often vast networks of compromised computers, known as botnets. Each individual source might not be powerful enough to disrupt a system, but collectively, they generate a flood of traffic that can incapacitate even the most robust infrastructure.

There are different forms of DDoS attacks, each designed to target a specific aspect of a system's functionality. Volume-based attacks, for example, aim to saturate a network's bandwidth, while protocol attacks target network resources. Application-layer attacks, on the other hand, focus on specific applications, draining their resources and causing service interruptions.

What makes DDoS attacks particularly disruptive is their ability to disrupt not just a single target, but an entire network of connected systems. A successful attack can lead to severe downtime, potentially causing significant financial and reputational damage.

Moreover, DDoS attacks are often used as a smokescreen, diverting attention away from other malicious activities. While security teams scramble to mitigate the DDoS attack, cybercriminals might exploit the chaos to infiltrate systems, exfiltrate data, or launch other forms of attacks.

As the digital landscape evolves and our reliance on connectivity increases, the potential impact of DDoS attacks grows as well. These attacks, once the purview of seasoned hackers, are now available as a service on the dark web, making it even more crucial for us to have effective measures in place to prevent and mitigate such onslaughts.

The importance of safeguarding our systems against DDoS attacks cannot be understated. In the next sections, we will delve into the mechanisms behind these attacks and explore effective strategies to protect against them.

2.2.3.1. How DDoS Attacks Work

To understand how Distributed Denial of Service (DDoS) attacks work, we must first view them not as single incidents, but as orchestrated campaigns of chaos. They are akin to an orchestra of rogue instruments; each playing its part to overwhelm the senses and drown out the melody of regular network traffic.

In a DDoS attack, the attackers begin by building an army of remote-controlled computers or 'bots', forming a network known as a 'botnet'. This is typically achieved through various forms of malware, which infect the system and allow the attacker to control it remotely. Each infected system, or 'bot', can then be used to generate a flood of network traffic, directed towards the target.

The key to a DDoS attack lies in its distributed nature. The attack doesn't come from one source but from many, often thousands or even millions. These sources can be scattered across the globe, making it challenging to identify and block the attack's origin. It's like trying to stop an army of ants, each insignificant on its own, but formidable as a group.

The targets of DDoS attacks are typically servers or networks that host websites or provide other online services. When the attack is launched, the targeted system is inundated with requests, consuming its bandwidth and processing capacity. Like a highway choked with traffic, the targeted system becomes congested, slowing to a crawl or even grinding to a halt.

The specific tactics used can vary, but the goal is the same: overwhelm the target until it can no longer provide the service for which it is designed. Some DDoS attacks focus on consuming bandwidth (volume-based), some exploit weaknesses in network protocols (protocol-based), and others target the application layer, attempting to exhaust system resources like processing power and memory (application layer attacks).

2.2.3.2. Mitigating DDoS Attacks

Defending against a Distributed Denial of Service (DDoS) attack is akin to a strategic dance, a choreographed response to an unpredictable and chaotic adversary. It involves a blend of preparedness, rapid detection, and swift, coordinated countermeasures.

Initiating the defensive choreography is the design of a robust infrastructure, one that inherently resists DDoS attacks. Redundancy is a cornerstone of this design, creating multiple pathways for network traffic and avoiding single points of failure. Load balancing across servers and data centres can help manage the sudden influx of traffic during an attack, reducing the chance of system overload.

Furthermore, bandwidth overprovisioning can offer a buffer against volume-based attacks, while implementing secure network protocols can help safeguard against protocol-based ones. A system fortified in such a manner presents a hardened target, capable of enduring the initial wave of an assault.

Early detection is the second critical step in this dance. Employing real-time network monitoring tools allows for swift identification of abnormal traffic

61

patterns, serving as an early warning system for a potential DDoS attack. Coupled with this is the use of artificial intelligence and machine learning algorithms that can detect subtle anomalies and predict attack vectors, providing a proactive defence mechanism.

Once an attack is detected, the immediate countermove is to filter and reroute the malicious traffic. This is where DDoS mitigation services come into play. These services, working in conjunction with Internet Service Providers (ISPs), can segregate normal traffic from the attack traffic, steering the latter away from the target and into a sort of digital cul-de-sac where it can be neutralised.

Further tactics include rate limiting, which restricts the number of requests a server will accept within a certain time frame, and IP blacklisting, blocking traffic from known malicious sources.

And finally, collaboration plays an essential role in this defensive dance. Sharing information about DDoS attacks can help the broader internet community develop more effective mitigation strategies. It's an acknowledgment that DDoS attacks are not just a threat to individual organisations, but to the global digital ecosystem as a whole.

Mitigating DDoS attacks is a continuous, evolving process. As the tactics of attackers shift and adapt, so too must the defences. It's a dance that never ends, a perpetual manoeuvring to protect the accessibility and integrity of our digital world.

2.2.4. Insider Threats

Beyond the barricades of firewalls and beneath the watchful eyes of intrusion detection systems, lurks a different breed of threat - the insider.

These individuals, due to their access and understanding of the organisation's systems, pose a unique and potentially devastating cyber risk.

Insider threats can manifest in numerous ways. An employee with malicious intent, the disgruntled programmer or spiteful manager, may exploit their access to sensitive data for personal gain or to inflict harm on the organisation. They have the potential to bypass security measures from the inside, making their actions particularly hard to detect and prevent.

Alternatively, a well-intentioned but negligent employee could inadvertently expose the organisation to cyber threats. Such unintentional insider threats often arise from poor cybersecurity practices, such as weak password use, clicking on malicious links, or misconfiguring security settings. While lacking malice, the consequences of these actions can be just as damaging.

Then there are the infiltrators. External actors who have gained access to the system through means such as social engineering, phishing or other forms of deception. They masquerade as legitimate users, all the while stealing data or causing disruption.

Managing insider threats necessitates a balanced approach that blends technology, processes, and people. On the technological front, solutions like Data Loss Prevention (DLP) tools, user behaviour analytics, and access control mechanisms can detect and mitigate the risk of insider threats.

Well-defined processes play an essential role too. Regular audits of user activities, especially those with privileged access, can help identify suspicious behaviour. Similarly, procedures should be in place for managing the digital access of departing employees, ensuring that their privileges are promptly revoked.

Lastly, a culture of cybersecurity awareness is paramount. Regular training programs can equip employees with the knowledge to avoid common cyber pitfalls and to report suspicious activities. It encourages them to be part of the solution, rather than being potential vulnerabilities.

In essence, combating insider threats is as much about managing people and processes as it is about deploying the right technology. It's about nurturing a trusted, security-conscious environment that can both benefit from, and protect, the openness and connectivity of the digital world.

2.2.4.1. *Malicious vs Unintentional Insider Threats*

The realm of insider threats is not monolithic. There are subtle nuances that distinguish different types of insider threats, especially when it comes to intent. Here, we delve into the contrasting natures of malicious and unintentional insider threats, two contrasting forces that nevertheless pose substantial risks to cybersecurity.

Malicious insider threats are marked by deliberate intent to harm. These are individuals within an organisation who, driven by motivations such as personal gain, revenge, or ideological reasons, consciously seek to exploit their access and knowledge of the systems. They deliberately target sensitive data, critical infrastructure, or key operational processes, aiming to steal, alter, or disrupt. The very purpose of their actions is to inflict harm upon the organisation.

The cloak of trust and legitimacy that insiders wear makes detecting and mitigating these threats particularly challenging. They operate from within the established security perimeters, effectively hiding in plain sight. Their intimate knowledge of the system's layout and operations can enable them to navigate covertly, often leaving minimal traces of their malevolent activities.

On the other end of the spectrum lie unintentional insider threats. These individuals, unlike their malicious counterparts, bear no ill intent. They are not villains but victims themselves, ensnared by crafty phishing campaigns, seduced by cleverly disguised malware, or simply caught off guard due to poor cybersecurity hygiene.

Despite their lack of malice, the cybersecurity implications of unintentional insider threats are by no means less serious. Inadvertent data leaks, unintentional enabling of ransomware attacks, or accidental exposure of system vulnerabilities can inflict substantial damage to an organisation.

Frustratingly, the benign nature of these threats often makes them more challenging to address. After all, how does one guard against good intentions gone wrong? Mitigating these risks requires an approach that emphasises education, awareness, and a culture of shared responsibility towards cybersecurity.

In conclusion, understanding the dichotomy between malicious and unintentional insider threats is crucial. It highlights the need for a diverse, nuanced approach to insider threat management – one that incorporates robust technological safeguards, thorough procedural controls, and a strong cybersecurity culture. It is only through such an integrated approach that organisations can effectively safeguard their cyberspace from the dangers lurking within.

2.2.4.2. Strategies to Mitigate Insider Threats

Addressing the potential risk from within requires a thoughtful, layered approach. Insider threats, be they accidental or malicious, demand a blend of technological, procedural, and cultural strategies to curtail their potential damage. Here are key strategies that can form the bulwark against insider threats.

Foremost, organisations must prioritise user education and awareness. The human element is often the weakest link in cybersecurity, but it can also be its strongest ally. Regular training sessions on cybersecurity hygiene, clear communication about acceptable use policies, and drills simulating phishing and other attacks can help employees recognise and avoid potential threats.

Access management is another critical factor. Applying the principle of least privilege, i.e., providing employees only the access they need to perform their duties, can limit potential damage. Regular audits to review user privileges can help identify and rectify excessive access rights.

Employing User and Entity Behaviour Analytics (UEBA) can also aid in spotting anomalous behaviour. By studying regular usage patterns, these systems can identify unusual activities that may signify a threat, enabling swift action. Similarly, Data Loss Prevention (DLP) tools can monitor and control data movement within the network, helping prevent unauthorised data transfers.

Establishing a robust incident response plan is another crucial step. Such a plan outlines procedures to follow when a security incident is detected, helping ensure swift and effective action that minimises potential harm.

Institutionalising a positive security culture is equally vital. Encouraging employees to voice their concerns, rewarding those who identify security risks, and fostering an environment where security is seen as a shared responsibility can greatly enhance an organisation's cybersecurity posture.

Lastly, it's essential to maintain open lines of communication between IT, HR, and management. Personnel changes, grievances, or other issues could indicate potential threats that need to be addressed proactively.

In summary, mitigating insider threats demands a layered, comprehensive approach. By combining technological tools with procedural controls and a positive security culture, organisations can significantly strengthen their defence against threats originating from within their own ranks.

2.3. Advanced Persistent Threats (APTs)

In the murky depths of the cyber threat landscape lurk menaces that combine audacity with patience, technical prowess with strategic finesse. These are the Advanced Persistent Threats, or APTs. As their name suggests, APTs aren't your run-of-the-mill opportunistic attacks. Instead, they represent an intricate blend of persistence, sophistication, and usually, state-level sponsorship.

The primary distinguishing feature of APTs is their longevity. Unlike most cyberattacks that seek immediate gratification, APTs play a long game. Their objective is often espionage or sabotage, and the targeted assets usually include intellectual property, strategic data, or critical infrastructure. The perpetrators—often state-sponsored cyber warfare units or highly organised crime syndicates—are adept at maintaining their presence undetected within compromised systems for extended periods.

The seriousness of the APT menace derives from its remarkable resilience and adaptability. APTs are designed to penetrate deeply, spread widely, and stay rooted against attempts at elimination. They showcase a high degree of technical sophistication and employ a blend of techniques, including custom malware, zero-day vulnerabilities, and social engineering.

In the following sections, we delve into the intricate mechanics of APTs, explore some of the notorious APT groups, and present strategies for defending against this persistent menace. This knowledge will illuminate the sophistication of the threats that organisations face in the digital arena and underscore the need for a comprehensive, agile, and multi-faceted cybersecurity approach.

2.3.1. Understanding APT Lifecycle

An Advanced Persistent Threat (APT) is akin to a meticulously planned and executed heist. It's not a smash-and-grab operation but rather a covert infiltration, conducted with the patience of a stalker and the precision of a surgeon. To truly grasp the nuances of APTs, it's imperative to comprehend their lifecycle—a series of steps designed to compromise a target while remaining undetected for as long as necessary.

1. **Reconnaissance**: The APT lifecycle begins with thorough research about the potential target. The perpetrators gather information about the target's digital footprint, security posture, employees, partners, and business operations. They use a mix of open-source intelligence (OSINT), social engineering, and possibly even physical espionage.

2. **Incursion**: Once the perpetrators have gathered sufficient data, they design an entry strategy. This typically involves exploiting known vulnerabilities, using zero-day attacks, or employing social engineering techniques. Regardless of the method, the goal is to penetrate the target's defences without raising alarms.

3. **Discovery**: Post-incursion, the APT begins exploring the compromised system, identifying its structure, locating valuable data, and recognising security measures. The goal is to map the environment for potential expansion and data exfiltration.

4. **Capture**: Once the APT identifies valuable data, it captures the information. This could involve creating copies of data files, intercepting network traffic, recording keystrokes, or taking screenshots.

5. **Exfiltration**: The captured data is then exfiltrated to the attackers' servers. The perpetrators employ a variety of tactics to remain undetected, including encrypting the data, using covert channels, or even manipulating the data's timing and volume to blend with regular network traffic.

6. **Maintenance**: The hallmark of APTs is their persistence. The attackers strive to maintain their presence within the compromised

69

systems, deploying additional backdoors, creating redundant exfiltration paths, and continually updating their tools to evade detection.

7. **Expansion**: Often, the APT will attempt to expand its presence within the network, compromising additional systems, escalating privileges, and further entrenching its presence.

By examining the lifecycle of APTs, it's evident that these are not impulsive attacks but systematic campaigns. The perpetrators are patient, methodical, and innovative traits that make APTs particularly challenging to detect and eliminate. In the upcoming sections, we will examine strategies to tackle this nuanced menace.

2.3.2. Case Studies of APT Attacks

APT attacks bear a disconcerting reality, illustrating the cunning and persistence of threat actors, and their ability to disrupt even the most fortified systems. Let's explore a few notable instances that underscore their far-reaching implications.

Operation Aurora - Uncovered in 2010, Operation Aurora aimed at numerous high-profile organisations, including Google, Adobe, and Juniper Networks. The attackers exploited a zero-day vulnerability in Internet Explorer and used spear-phishing emails to gain access. The assault intended to gain control over sensitive systems and steal intellectual property. The intricacy, precision, and persistence led analysts to believe that it was a state-sponsored attack.

Stuxnet - Unveiled in 2010, Stuxnet was a malicious computer worm that disrupted Iran's nuclear program. The worm targeted the programmable logic controllers (PLCs) used to automate industrial processes. Stuxnet's ability to cause physical damage to infrastructure marked a new era in cyber warfare. Analysts widely regard it as a joint effort between American and Israeli intelligence agencies.

APT28 (Fancy Bear) - Active since the mid-2000s, APT28 is believed to be associated with the Russian government. The group has conducted several high-profile attacks, including the 2016 U.S. Democratic National

Committee (DNC) breach. APT28 often uses spear-phishing emails and watering hole attacks to compromise its targets, and it has a penchant for targeting governmental, military, and security organisations.

APT29 (Cozy Bear) - Another group linked to the Russian government, APT29, is notorious for its extensive, long-term campaigns. The group uses highly customised tools and has a preference for fresh, zero-day exploits. APT29 is allegedly responsible for the 2014 White House breach and the 2015 State Department breach.

APT10 (MenuPass Group) - Linked with the Chinese government, APT10 is known for its relentless campaigns against organisations in various sectors, including healthcare, aerospace, and telecommunications. The group often uses spear-phishing emails, custom malware, and living-off-the-land tactics.

These case studies paint a vivid picture of APT attacks' scale, sophistication, and potential damage. They underline the need for advanced threat detection systems, well-trained cybersecurity personnel, and robust incident response plans.

2.4. Case Studies of Notable Cyber Attacks

In an increasingly digital world, cyber-attacks have become common headlines, often exposing millions of users' data, causing extensive downtime, and resulting in financial loss. These stories serve not merely as cautionary tales, but also as learning opportunities. Studying these high-profile cyber-attacks helps shed light on the strategies used by threat actors, the vulnerabilities they exploit, and the measures organisations can take to bolster their defences. From data breaches and ransomware attacks to state-sponsored hacks, this section unravels some of the most impactful cyber-attacks and their reverberating effects on global cybersecurity.

2.4.1. NotPetya: The Most Destructive Cyber Attack

The NotPetya cyber-attack in 2017, widely regarded as one of the most destructive in history, shows the dire consequences of a well-orchestrated and malicious cyber operation. Originating in Ukraine and quickly spreading across the globe, this cyber onslaught left companies staggering under millions of dollars in damages and tested their resilience in unprecedented ways.

Unlike typical ransomware that seeks financial gain, NotPetya was designed with a far more destructive intent - data annihilation. This wiper malware, disguised as ransomware, targeted Windows-based systems, exploiting multiple vulnerabilities to encrypt the master boot record, causing systems to crash and rendering data irretrievable. As such, even if victims opted to pay the demanded ransom, data recovery was impossible.

The rapid proliferation and extensive damage caused by NotPetya underscores the perilous reality of sophisticated malware. Moreover, the incident demonstrated that no organisation is immune to such threats, with victims spanning diverse industries and regions, including shipping giant Maersk, pharmaceutical manufacturer Merck, and numerous other firms and institutions.

The NotPetya attack served as a global wake-up call, emphasising the need for advanced, multi-layered cybersecurity measures, timely software updates, and robust disaster recovery plans. It highlighted the potential for cyber warfare to cause devastating effects on a global scale and set a new precedent for the destructive potential of cyber-attacks.

In the aftermath of NotPetya, cybersecurity policies worldwide underwent drastic reviews, organisations ramped up their IT infrastructure investments, and there was an increased focus on incident response and cyber threat intelligence sharing. The event not only altered the threat landscape but also fundamentally reshaped our understanding of how we approach and manage cybersecurity.

2.4.2. Equifax Data Breach: A Case of Unpatched Vulnerability

The story of the Equifax data breach offers a harsh lesson on the importance of timely system updates and vulnerability management. Occurring in 2017, the breach saw one of the largest ever leaks of personal data, compromising the sensitive information of approximately 147 million people. This breach was not the result of an exceedingly complex or unknown attack method. Instead, it was due to the negligence of patching a known vulnerability.

Equifax, one of the largest consumer credit reporting agencies, fell prey to a simple web application vulnerability in Apache Struts, an open-source framework for developing Java applications. An available patch for this vulnerability had been released months before the breach, but Equifax failed to implement it. This lapse in basic cybersecurity hygiene allowed attackers to access their systems and exfiltrate an enormous amount of sensitive data, including names, social security numbers, birth dates, addresses, and in some cases, driver's license numbers and credit card numbers.

The fallout from the breach was massive, demonstrating the cascading consequences of a single unpatched vulnerability. Equifax's stock price plummeted, lawsuits flooded in, and the company had to spend millions on response efforts and credit protection services for affected customers. Furthermore, it led to the resignation of the CEO and severely tarnished the company's reputation.

The Equifax breach underscores the significance of maintaining a robust patch management system and the disastrous effects of complacency in this area. It also emphasised the need for continuous vulnerability scanning and stringent oversight of cybersecurity practices, especially in organisations handling vast amounts of sensitive data. This event is a powerful reminder that the cost of prevention is often far less than the cost of a breach.

2.5. The Dark Web and Cybercrime

Picture a city at night, with bustling activity in the bright, well-lit areas. The public squares are filled with people engaged in legitimate, daily life activities. Then there are the shadowy alleyways, where less savoury activities occur, far from the prying eyes of law enforcement. In the digital world, the Dark Web is equivalent to these alleyways.

The Dark Web is a fascinating and disturbing realm of the internet, often associated with illegal activities, including cybercrime. It is a part of the internet that isn't indexed by traditional search engines and can only be accessed using special software, such as Tor or I2P, which allow users and website operators to remain anonymous or untraceable.

As we delve deeper into understanding cyber threats and attacks, it becomes critical to shed light on this elusive part of the internet. It is here that cybercriminals often conduct their transactions and trade their ill-gotten goods. It's also a hub for selling hacking tools and services, which can be used by individuals or groups to launch cyber-attacks. Understanding the Dark Web and its relation to cybercrime is instrumental in gaining a holistic perspective of the cyber threat landscape.

2.5.1. Understanding the Structure of the Dark Web

Envision an iceberg, its majestic peak merely a fragment of its entire form. Much like this iceberg, the internet, as most people know it, is only a small portion of what's actually out there. The Dark Web represents the hidden mass beneath the waterline, part of a colossal digital construct rarely accessed by everyday users.

The internet is broadly divided into three layers: the Surface Web, the Deep Web, and the Dark Web. Each represents a segment of the web's broader structure and varies in terms of accessibility and content.

The Surface Web is the part of the internet that most of us interact with daily. It comprises websites and resources that are open to the public and can be found using regular search engines like Google.

A step deeper lies the Deep Web. Here, you'll find pages and databases that are not indexed by standard search engines but are still mostly legal and safe. The Deep Web includes resources like private databases, restricted website content, and pages behind paywalls or passwords.

Finally, hidden within the lowest recesses of the internet structure is the Dark Web. This segment of the internet is deliberately hidden and is only accessible through specific anonymising software such as Tor (The Onion Router) or I2P (Invisible Internet Project). It hosts an array of websites that are anonymised and untraceable, hence a haven for illicit activities, including cybercrime.

The Dark Web's structure fosters an environment of anonymity. It is built upon an overlay network, which runs on top of the regular internet. This layering is made possible using encryption tools, which ensure that a user's location and identity remain concealed. Due to these characteristics, the Dark Web has become a hub for illicit activities, fostering an economy of cybercrime that poses significant challenges to cybersecurity.

2.5.2. Cybercrime Markets and Forums on the Dark Web

Sweeping the dust from the obscurity, it becomes evident that the Dark Web harbours an underworld marketplace. A place where masked identities conduct transactions with a disturbing degree of professionalism and customer service. These markets are the go-to venues for various forms of cybercrime, from the trading of stolen data to the hiring of hacker services.

Much like traditional e-commerce platforms, these darknet markets offer a vast array of illegal products and services. They range from stolen credit card data, hacked account credentials, and personal identification information, to malicious software such as ransomware, and even services like hacking-for-hire. All transactions typically occur through untraceable cryptocurrencies, predominantly Bitcoin, maintaining the buyer's and seller's anonymity.

Forums on the Dark Web further provide an arena where cybercriminals can gather, discuss, and collaborate. These forums serve multiple purposes. They are spaces for hackers to share techniques, sell unique malware or exploits, and even provide mentorship to novice hackers. Topics can vary greatly, from the rudiments of script execution to complex zero-day exploits and advanced persistent threats (APTs).

Understanding these marketplaces and forums is crucial for cybersecurity. They not only present a clear and present danger but also provide insight into the tactics, techniques, and procedures used by cybercriminals. Tracking these forums, understanding the evolving language and slang, and staying up to date with discussed topics can serve as an early warning system for emerging threats, allowing cybersecurity professionals to better anticipate and prepare for new types of cyber-attacks.

Despite its ominous aura, the Dark Web is not an all powerful, unassailable fortress. Successful law enforcement operations have managed to infiltrate and shut down many of these criminal marketplaces, resulting in arrests and significant disruptions to the cybercrime economy. However, the adaptive nature of these spaces and the growing demand for illicit goods and services imply that cybersecurity efforts need to remain dynamic and persistent to protect against the evolving threats that originate from the Dark Web.

2.5.3. Law Enforcement Challenges and Strategies

Navigating the labyrinthine dark alleys of the Dark Web poses significant challenges for law enforcement agencies worldwide. The invisible barriers, constructed by intricate layers of anonymisation and encryption, cloak cybercriminal activities and hinder the pursuit of justice.

A key challenge lies in the inherent anonymity provided by the Dark Web. Users can access websites while remaining largely untraceable due to the layering of IP addresses provided by encryption tools such as TOR (The Onion Router). This obscuration makes it difficult for law enforcement to identify, track, and apprehend cybercriminals who exploit the Dark Web's anonymous nature.

In addition to anonymity, jurisdictional limitations further impede the fight against cybercrime. With the Dark Web's global reach, cybercriminals from one country can easily target victims in another. This raises complex jurisdictional issues, as law enforcement agencies must navigate international laws and cooperate with foreign counterparts to investigate and prosecute these crimes effectively.

On the bright side, the complex challenges do not render law enforcement powerless. A range of strategies has been employed to uncover the illicit activities concealed in this cyber netherworld.

Law enforcement agencies use advanced cyber forensics and sophisticated investigative techniques to unmask Dark Web users. These methods include traffic correlation attacks, de-anonymisation attacks, and the exploitation of software vulnerabilities. Coupled with traditional investigative techniques, these tools help to pinpoint and gather evidence against cybercriminals.

Also, international cooperation plays a crucial role in combating Dark Web crimes. Agencies worldwide are increasingly collaborating, sharing information and resources to coordinate operations. Some notable successes include the takedown of the infamous darknet marketplaces like Silk Road and AlphaBay.

Another strategy revolves around infiltrating darknet markets and forums. Undercover operations can unmask individuals involved in illegal activities, disrupt criminal operations, and gather valuable intelligence. This approach requires agents to possess deep knowledge of the cybercriminal landscape, along with specialised technical skills.

While these strategies have led to significant victories, the game of cat-and-mouse between law enforcement and cybercriminals persists. Continuous development and refinement of investigative tools, enhancement of international cooperation, and investment in personnel training are all imperative for maintaining the upper hand in this ongoing battle against cybercrime on the Dark Web.

CHAPTER 3. CYBERSECURITY PRINCIPLES AND CONCEPTS

Woven into the tapestry of the digital world, cybersecurity is not just an array of countermeasures, but a discipline founded on enduring principles and concepts. This chapter unveils the underlying structure of cybersecurity - the guiding ideas and paradigms that enable us to combat an ever-evolving threat landscape.

At the heart of the cybersecurity field are its principles - the foundational concepts that shape strategies and guide practices. They serve as a lighthouse, illuminating the way forward amidst the fog of a complex cyber threat landscape. From understanding the principles of least privilege to embracing defence in depth, these guiding tenets offer a way to navigate the intricate world of cybersecurity.

However, cybersecurity is not just about defending against threats. It also encompasses the protection and preservation of key attributes of information - confidentiality, integrity, and availability. Often bundled together as the CIA triad, these elements form a cornerstone of cybersecurity, guiding professionals in creating strategies that ensure the reliable, accurate, and timely access to data.

Beyond principles and the CIA triad, this chapter will introduce the concept of risk management in cybersecurity. A delicate dance of identification, evaluation, and mitigation of risks, this process allows an organisation to align its cybersecurity strategies with its overall business objectives.

Finally, no cybersecurity journey would be complete without understanding the human element. Recognising that cybersecurity isn't just a technical issue, but also a human one, is critical. This chapter will explore how an organisation can build a robust cybersecurity culture that reinforces secure behaviours and attitudes.

From laying down the foundational principles to understanding key concepts such as the CIA triad and risk management, this chapter offers a comprehensive guide to the pillars of cybersecurity. With the human element as the keystone, it aims to weave together a holistic view of cybersecurity, providing readers with a strong foundation upon which to build their knowledge and skills.

3.1. Understanding the CIA Triad

Confidentiality, Integrity, and Availability, abbreviated as the CIA Triad, form the very bedrock of information security. Together, they guide the way we protect and manage information, shaping the policies and controls used in cybersecurity. Understanding the principles of the CIA Triad, and their implications, is vital to any discourse in cybersecurity.

All three principles of the CIA Triad are interrelated and balancing the importance of each, depending on the context, is a continuous task for cybersecurity professionals. Too much emphasis on confidentiality and integrity might hinder availability, whereas focusing on availability might expose sensitive data to unauthorised access.

The CIA triad serves as a simple yet powerful model for understanding and addressing the challenges of cybersecurity. It helps organisations identify and classify their information assets based on their need for confidentiality, integrity, and availability, enabling them to implement appropriate controls and measures for protection. By striving to uphold these three principles, we can navigate the turbulent seas of the digital world with confidence and assurance.

3.1.1. Confidentiality: The First Pillar of Security

Confidentiality carries profound significance in the realm of cybersecurity. At its essence, it safeguards sensitive information from falling into unauthorised hands, creating a veil of secrecy around what needs to be kept hidden. A breach of confidentiality isn't merely a technical failure; it's a betrayal of trust, often with far-reaching implications.

The significance of confidentiality is underscored by the diverse array of information that requires protection. Medical records, personal identifiers, financial data, trade secrets, and national security intelligence are all subject to confidentiality. In each of these cases, unauthorised access can lead to catastrophic outcomes, from identity theft to corporate espionage, and even threats to national security.

To uphold confidentiality, various technologies and strategies are deployed. One common method is encryption, a mathematical algorithm that obscures information, rendering it unreadable to anyone who lacks the decryption key. Confidentiality can also be maintained through access controls, which restrict data access to authorised users. This involves a careful combination of hardware, software, and procedural methods, such as multi-factor authentication.

While the implementation methods may vary, the goal remains the same: to restrict information access to the right individuals, ensuring that the information remains confidential. Confidentiality, therefore, is more than a cybersecurity principle; it's a commitment to protecting what is valuable and maintaining trust in an era where data is the new gold.

While ensuring confidentiality, we must also consider its interplay with the two other pillars of security: integrity and availability. Striking a balance between the three is vital, and often challenging, task for cybersecurity professionals. Regardless, confidentiality remains the cornerstone of cybersecurity, a guardian of secrets in a world that is constantly under the siege of cyber threats.

3.1.2. Integrity: Ensuring Data Accuracy and Reliability

The second pillar of the CIA Triad, integrity, speaks volumes about the robustness and reliability of data. It safeguards the authenticity of data, ensuring it remains untampered and uncorrupted throughout its lifecycle.

Imagine a world where bank statements could be manipulated, medical records falsified, or software code maliciously altered. The chaos ensuing from such situations underscores the critical role of integrity in cybersecurity. We rely on the certainty that data, once created or stored, can only be altered by authorised entities and in sanctioned ways.

Ensuring data integrity begins with user authentication, verifying that individuals are who they claim to be. This step validates that only authorised users can make changes to data. Beyond user authentication, change management processes monitor and control how data is altered, often keeping an audit trail to track who made changes, when, and why.

But what happens if data integrity is breached? Hashing is a cryptographic technique used to detect changes to data. By taking an input and producing a fixed-size string of characters, hashing identifies even minuscule changes, similar to a data fingerprint. Should the fingerprint not match, it indicates a breach of integrity.

From a broader perspective, integrity extends to entire systems, safeguarding against unauthorised activities that could disrupt services. This is the essence of system integrity, an extension of the integrity principle to ensure the robustness and availability of entire systems.

In the realm of cybersecurity, maintaining integrity is like being a sentry guarding a fortress. It's about ensuring that the walls remain strong, the gates unbreached, and the people within safe. As we continue to traverse the digital landscape, integrity stands as a stalwart guardian, upholding the authenticity and trustworthiness of our data and systems.

3.1.3. Availability: Keeping Systems Accessible

Availability, the third arm of the CIA triad, focuses on the accessible and timely delivery of information and resources to authorised entities when needed. Just as a locked library hampers a scholar's research, lack of access to digital data could stifle a business's operations, make crucial healthcare records unreachable, or bring governmental services to a halt.

Under the tenet of availability, data is as useful as it is accessible. Cybersecurity measures work to ensure that authorised users have constant, uninterrupted access to resources and services. This involves both keeping systems running smoothly and preventing disruption due to hardware failures, system upgrades, or cyberattacks.

In the spirit of ensuring availability, data redundancy strategies, like backup and replication, are used. These processes make copies of critical data and store them in multiple locations, safeguarding against loss or corruption of the original. Meanwhile, techniques like failover and load balancing help ensure the availability of services by distributing network traffic or reallocating resources when a server or system fails.

Yet, availability is not just about hardware and system design; it also deals with network security. Cybersecurity professionals use tools to monitor networks and detect threats like Denial-of-Service (DoS) or Distributed Denial of Service (DDoS) attacks that overwhelm a network and make it inaccessible.

The essence of availability transcends the uninterrupted performance of systems. It mirrors the modern-day expectation of 24/7 accessibility. With the internet embedding itself into every aspect of life, ensuring the availability of systems and services is more important than ever. Like a well-oiled machine, the digital world runs smoothly when its parts are available, upholding the rhythm of this ceaseless dance.

In cybersecurity, the principles of Confidentiality, Integrity, and Availability stand as pillars supporting the secure flow of digital life, each contributing uniquely to the protection and sustenance of our cyber universe.

3.2. The Principle of Least Privilege

Have you ever pondered the wisdom in the phrase, "Don't put all your eggs in one basket?" This saying encapsulates the essence of the Principle of Least Privilege (PoLP), a critical concept in cybersecurity that regulates access rights for users, applications, systems, and processes within a network. This section peels back the layers of this principle, revealing how it acts as a gatekeeper, ensuring that access to sensitive data and functions is as limited as it is necessary.

In the grand tapestry of digital connections, threads of data interweave to form an intricate picture of information flow. However, not all threads should have equal access to the entire design. The Principle of Least Privilege posits that a user or program should only be given the minimal access – or privileges – required to complete its function, nothing more.

This approach not only decreases the potential surface for attacks but also minimises the possible damage in case of a breach or internal misuse of access rights. Like a well-organised assembly line, each participant in the digital dance has a specific role and is empowered to perform only within that domain. Stay with us, as we explore the mechanics, implementations, and importance of the Principle of Least Privilege in cybersecurity.

3.2.1. Definition and Benefits of the Principle of Least Privilege

The Principle of Least Privilege (PoLP) is a cybersecurity guideline that advises limiting access rights for users, applications, systems, and processes to the bare minimum they need to execute their designated tasks. The underlying logic of this principle is to curb the potential damage caused by the misuse or compromise of these privileges. Like a finely curated access pass, it only grants the essentials, thus keeping excess privileges in check and risk under control.

Visualise a bank vault, where every employee has a specific role, yet not every worker has access to all assets within the vault. A cashier may not need the same level of access as a bank manager. The Principle of Least Privilege operates similarly within digital systems, providing the user with just enough access to perform their function, thus reducing the potential blast radius if that user's account becomes compromised.

The benefits of employing the Principle of Least Privilege are manifold. For one, it minimises the risk of internal and external data breaches by reducing the number of opportunities for attackers to gain unauthorised access to sensitive information. It also simplifies the task of managing user permissions and maintaining system stability. In the event of an attack, the implementation of PoLP makes it easier to track and rectify any unauthorised changes to the system.

Moreover, PoLP can limit the damage resulting from inadvertent errors. An employee with broad system access might, through a simple mistake, modify or delete important system files, causing extensive damage. By ensuring employees only have the access they need, such accidents are less likely to occur.

Finally, regulatory compliance, a recurrent theme in cybersecurity, benefits from the Principle of Least Privilege. Numerous regulations stipulate limiting user access, and abiding by the PoLP assists organisations in meeting these requirements, dodging penalties, and maintaining their reputation.

3.2.2. Implementing Least Privilege in Practice

Enacting the Principle of Least Privilege might appear as a simple undertaking on the surface, but in practice, it requires diligent planning, continuous monitoring, and regular adjustments to keep pace with changing roles, tasks, and technological advancements.

Starting from the inception of user accounts, the Principle of Least Privilege comes into play. New accounts should be created with the minimum necessary access rights, limiting the potential avenues for exploitation right from the get-go. However, defining what "minimum necessary access" means for each role can be a challenging task. It involves having a comprehensive understanding of various job functions and their corresponding access requirements.

Once roles and corresponding privileges have been established, a robust system of access control can be put in place. These controls could range from basic password protection to more advanced measures such as multi-factor authentication and biometric verification. Organisations might also use role-based access control (RBAC) systems, which tie permissions directly to job functions rather than individuals, allowing for efficient control over user privileges.

Another critical aspect of implementing PoLP is periodic auditing and review of access privileges. This process involves identifying any discrepancies or excessive privileges and adjusting them as needed. If a user changes roles within the organisation, their privileges should be reassessed and modified to reflect their new responsibilities. This sort of periodic reassessment can prevent privilege creep - the gradual accumulation of unnecessary access privileges over time.

Yet another layer of safeguard is logging and monitoring activities, which can provide insights into unusual user behaviour or suspicious patterns. Anomalies can be swiftly detected, and potential security incidents could be proactively stopped in their tracks. It is crucial to remember that even with minimal privileges, users can pose potential security risks.

In this era of digital transformation, where organisations are moving more and more resources to the cloud, the Principle of Least Privilege finds a new stage. Cloud platforms and software-as-a-service (SaaS) applications often come with fine-grained access control capabilities, allowing for meticulous implementation of PoLP. But, these platforms also come with their own set of challenges, including the need for specialised knowledge to correctly configure these controls and guard against misconfigurations.

Consideration should also be given to the software and applications that run within an organisation's network. Software should be designed and configured to operate with the least set of privileges necessary to complete its function, minimising the potential security risks in the event of a software vulnerability being exploited.

3.3. Defence in Depth: The Layered Approach to Security

Think of cybersecurity as an intricate game of chess. The game is not won by protecting the king with a single piece but rather by deploying a strategic, multi-layered defence that works together to safeguard the king. Similarly, ensuring robust cybersecurity is not about having one all-encompassing protection measure, but about having multiple layers of defence mechanisms working in unison. This strategy is known as Defence in Depth, a concept derived from a military strategy and adopted by the cybersecurity world.

Like an impregnable castle with numerous defence mechanisms, the Defence in Depth approach creates several barriers and checkpoints to delay, deter, and potentially thwart cybercriminals. This chapter will delve into this multi-faceted cybersecurity strategy, discussing its core principles, importance, and how it can be implemented to build a resilient digital defence fortress.

3.3.1. The Defence in Depth Strategy

Defence in Depth incorporates a wide array of security controls and mechanisms across various areas of an organisation's IT infrastructure. This includes physical security measures, such as secured access to data centres and server rooms, and technical controls, including firewalls, intrusion detection systems (IDS), intrusion prevention systems (IPS), and encryption methods.

Additionally, the strategy also considers the implementation of administrative controls like policies, procedures, and awareness training. No matter how fortified a system is, it's crucial not to overlook the human element, as user error or ignorance can often serve as an entry point for cyberattacks.

One of the critical benefits of the Defence in Depth strategy is that it doesn't rely on a single point of protection. If one layer is breached, the attacker is met with another layer of defence. This multiple barrier approach not only significantly slows down an attacker's progress but also increases the likelihood of detecting the attack in its early stages.

Moreover, the Defence in Depth strategy brings a level of redundancy to the security framework. Redundancy is an advantage in cybersecurity, contrary to many other aspects of business operations. If one security measure fails, others are in place as backups to keep the system secure. This greatly reduces the overall risk of a complete system compromise, enhancing the resilience of an organisation's cyber defence.

Implementation of a Defence in Depth strategy requires a comprehensive understanding of an organisation's IT infrastructure, including network topology, software applications, data flows, and operational processes. Once a complete overview is established, security layers can be strategically placed to protect critical assets and detect, slow, or halt any malicious activities.

However, the Defence in Depth strategy is not without its challenges. The complexity and cost associated with implementing multiple layers of security controls can be significant. It's also essential to manage the potential impact of these controls on system performance and user experience. These challenges highlight the need for a well-considered, carefully managed, and continuously reviewed approach to implementing a Defence in Depth strategy.

Overall, the Defence in Depth strategy is a fundamental principle in modern cybersecurity. When implemented effectively, it offers a robust, resilient, and comprehensive defence mechanism, creating a formidable barrier against cyber threats. As we continue to navigate through an increasingly digital era, characterised by sophisticated and evolving cyber threats, the relevance and importance of such a multi-layered defence approach become ever more apparent.

3.3.2. Applying the Layered Approach

As we delve into the application of the Defence in Depth strategy, it is essential to understand that the concept transcends the implementation of mere technical safeguards. It encapsulates a holistic approach that intertwines physical, technical, and administrative controls, uniquely tailored to meet an organisation's specific needs and risk appetite.

Physical Security Layer

The outermost shell of the layered defence is the physical security layer. It focuses on tangible, visible protection of assets. This involves the use of surveillance systems, biometric access controls, and secure lock mechanisms in data centres and offices to prevent unauthorised access. For instance, in a company housing servers on-premises, setting up access control systems and security cameras deters physical theft or tampering.

Network Security Layer

The next layer of defence is network security, aimed at protecting the data in transit across the network. Firewalls, Intrusion Prevention Systems (IPS), and secure network architecture design constitute this layer. For instance, implementing network segmentation to separate sensitive data from the general network can limit the extent of potential damage from a breach.

Host Security Layer

Host security focuses on protecting individual devices within the network such as servers, workstations, and mobile devices. Applying regular software patches and updates, using secure configurations, and installing antivirus software are part of this layer's strategy. An organisation, for example, can ensure that all its computers are set to automatically receive and install the latest security updates.

Application Security Layer

Application security safeguards the software applications that run on the host computers. It includes measures like input validation, secure coding practices, and regular vulnerability assessments. An e-commerce platform, for instance, might employ encryption techniques to protect user data and use penetration testing to identify and rectify security vulnerabilities.

Data Security Layer

At the heart of the Defence in Depth strategy is data security, a layer that protects the actual information that attackers typically target. Encryption of data both at rest and in transit, access controls, and data anonymisation techniques are used to protect sensitive data. For example, a healthcare organisation handling patient data could use encryption and strict access controls to protect this highly sensitive information.

User Awareness Layer

Last but not least, user awareness plays a crucial role in maintaining the overall security posture of an organisation. User training programs, regular phishing tests, and security awareness campaigns are essential components of this layer. A company might, for instance, regularly simulate phishing attacks to gauge employees' vigilance and provide them with feedback and further training.

By exemplifying the Defence in Depth strategy across different layers, organisations can create a comprehensive, multi-tiered approach to cybersecurity. While no single control can offer complete protection, the combination of diverse security measures across various layers significantly improves the overall security posture, reducing the likelihood and potential impact of a successful cyber-attack.

3.4. Risk Management in Cybersecurity

A cornerstone of cybersecurity is the effective management of risk. Just as a seafaring captain navigates treacherous waters with the help of a compass and map, the risk management process in cybersecurity allows organisations to identify, assess, and manage potential threats that could endanger their mission-critical assets.

Risk management is not about completely eliminating all risks - an impractical goal in a world where new vulnerabilities and threats emerge daily. Instead, it's about making informed decisions to control risk to an acceptable level, given an organisation's specific context and risk tolerance. It's an ongoing, iterative process that requires consistent monitoring, evaluation, and adjustment as the organisation's risk landscape evolves.

The risk management process typically comprises the following stages:

Risk Identification

The process begins by identifying potential threats and vulnerabilities that could adversely affect an organisation's information systems and data. Tools like threat intelligence feeds, vulnerability scanners, and risk assessment methodologies can help in this phase. It's not only about identifying external threats but also acknowledging internal vulnerabilities like weak access controls, outdated systems, or insufficient employee awareness.

Risk Analysis and Assessment

Once potential risks have been identified, the next step is to evaluate their potential impact and the probability of their occurrence. This step often involves qualitative and quantitative analysis. Qualitative analysis could include categorising risks as low, medium, or high, while quantitative analysis might involve assigning monetary values to potential losses from a risk.

Risk Mitigation

With the knowledge of what risks exist and their potential impact, organisations then decide how to address each risk. Choices include accepting the risk (if its impact is minimal), transferring the risk (through insurance, for example), mitigating the risk (by implementing controls to reduce its likelihood or impact), or avoiding the risk (by discontinuing the activity that introduces it).

Risk Monitoring and Review

Risk management is not a one-time event. It requires continuous monitoring and review to ensure that controls are effective and to identify any changes in the organisation's risk landscape. Regular audits, testing of controls, and reviews of incident response plans all contribute to this ongoing process.

Effective risk management goes beyond the IT department. It requires the involvement of stakeholders across the organisation, from frontline staff to senior executives, as cybersecurity is a shared responsibility that permeates every level of an organisation. A successful risk management program can help organisations optimise their security investments, achieve regulatory compliance, and foster trust among customers and partners, knowing that their data is handled responsibly and securely.

3.4.1. Risk Identification

As the initial step in the risk management process, risk identification forms the foundation upon which all other risk management activities are built. It provides a clear view of the hazards we face, shining a light on potential vulnerabilities and threats to the organisation's information systems and assets.

Risk identification involves a comprehensive examination of the organisation's environment to pinpoint elements that could jeopardise the security of its information assets. It is like a treasure hunt; except we're not looking for buried gold but potential perils that lurk beneath the surface.

The process extends beyond the walls of an organisation and encompasses both internal and external factors. Internally, this may include system vulnerabilities, lack of proper security measures, or weak security policies. Externally, it can involve any threat that could potentially exploit these internal vulnerabilities - from malicious hackers to natural disasters.

In the risk identification process, various tools and techniques come into play. Vulnerability scanners can probe systems to find weak spots, while threat intelligence feeds can provide information about the latest cyber threats. Risk assessment frameworks, such as the NIST's Risk Management Framework or the ISO 27005 standard, can provide a structured approach for identifying risks.

Risk identification also involves mapping assets and their value to the organisation. An asset can be tangible, like a server, or intangible, like proprietary data. By identifying these assets and understanding their worth, an organisation can better prioritise its risk management efforts. After all, a threat to a critical server housing sensitive customer data carries far more weight than a threat to a non-critical system with no confidential information.

It's important to note that risk identification is not a one-time event, but an ongoing task. As the organisation changes and grows, new vulnerabilities may arise, and new threats may emerge. Therefore, risk identification should be a regular activity, ingrained into the organisation's culture and operations.

At its core, risk identification sets the stage for informed decision-making. With a clear understanding of what threats exist and what vulnerabilities they might exploit, organisations can make strategic choices about where to focus their security resources and efforts, moving forward with a robust risk management strategy. In the stormy seas of cybersecurity, risk identification provides the compass that points the way to safer shores.

3.4.1.1. Understanding Risk Factors in Cybersecurity

In this ever-changing environment, understanding these risk factors allows organisations to stay ahead of potential threats and devise strategies to mitigate them.

Risk factors in cybersecurity are conditions or attributes that increase the potential for a security breach or the negative consequences of an attack. They can originate from an array of sources, varying from technical vulnerabilities in software, inadequate security policies, human error, or external threats, such as cybercriminals and nation-state actors.

Technical vulnerabilities often emerge from software bugs, misconfigurations, or outdated systems. They provide opportunities for threat actors to infiltrate an organisation's digital defences. As technology advances, so too does the sophistication of the tools threat actors use to exploit these vulnerabilities, underscoring the need for regular software updates and patches.

Yet, technology is but one facet of the cybersecurity landscape. The human element also plays a critical role. Inadequate training can leave employees susceptible to scams like phishing attacks, while the lack of a robust security culture can lead to careless handling of sensitive information. In essence, the human factor can often be the weakest link in the security chain.

An additional risk factor is the existence of inadequate security policies. Such policies may fail to address key areas of risk, lack enforcement, or be poorly communicated to staff. An effective security policy is comprehensive, clear, and properly implemented, providing guidelines that help reduce the risk of a security breach.

Lastly, external threats can significantly influence risk factors. The motives and capabilities of cybercriminals or state-sponsored attackers can dramatically alter the threat landscape. An organisation might face threats ranging from opportunistic hackers looking for easy targets to highly organised groups that carry out sophisticated, targeted attacks.

Risk factors are not standalone entities; they interact and overlap, often compounding the overall risk. For instance, a technical vulnerability paired with an untrained employee can result in a successful phishing attack. Therefore, understanding these risk factors is key to developing a comprehensive risk management strategy, one that addresses each component of risk and builds a resilient defence against cyber threats.

3.4.1.2. Methods for Identifying Risks

Numerous methods exist to identify risks, but this discourse will specifically delve into three of the most widely used: Vulnerability Assessment, Threat Modelling, and Automated Threat Intelligence.

Vulnerability Assessment is a systematic review of security weaknesses in an information system. It involves the identification and quantification of vulnerabilities in a system, application, or network. This process often employs specialised tools and software to scan for known vulnerabilities, such as outdated software, improper configurations, or lack of security controls. Once these vulnerabilities are identified, they can be catalogued, evaluated, and addressed, significantly reducing the likelihood of an attacker exploiting them.

Threat Modelling, on the other hand, is a method that focuses on potential threats or attacks against a system. It follows a structured approach, starting with the identification of system assets and followed by the comprehensive mapping of potential threats against these assets. This proactive approach helps in visualising the attack vectors, understanding the system's weak points, and subsequently fortifying them.

Automated Threat Intelligence, a newer entrant to the pantheon of cybersecurity strategies, leverages machine learning and artificial intelligence to automatically gather, analyse, and interpret data about potential threats. It scours a wide array of sources, including blogs, news sites, and even the dark web, looking for signs of new vulnerabilities or emerging threats. This real-time information allows organisations to stay up to date with the rapidly changing threat landscape and to take preventative measures before an attack happens.

Importantly, these methods should not be seen as independent processes, but rather as interconnected facets of a comprehensive risk identification strategy. The different methods complement each other, providing a 360-degree view of the organisation's security posture. The synthesis of this triad allows for an integrated approach that harnesses the strengths of each, thus enhancing the organisation's ability to anticipate, detect, and mitigate cybersecurity risks. It sets the sails right for the voyage through the tempestuous seas of cybersecurity, providing a beacon of guidance and foresight.

3.4.2. Risk Assessment

A deliberate, systematic process, risk assessment shapes the bedrock of effective cybersecurity management. It is a pivotal exercise that aids in discerning the risks that have been identified, assessing their potential impact, and prioritising remedial actions. By understanding what vulnerabilities are most critical, which assets are most valuable, and where the potential impacts of a breach are most severe, organisations can ensure they're applying their resources where they're most needed.

The primary task in a risk assessment process involves evaluating the severity of potential threats. This involves assigning a level of risk to each identified threat, based on a combination of the likelihood of occurrence and the potential impact. Likelihood represents the probability that a specific threat will exploit a particular vulnerability, while impact refers to the potential harm that could result from that exploitation. Both these factors are typically ranked on a scale, from low to high.

Consider, for instance, an outdated server software with known vulnerabilities that hasn't been updated. The risk assessment would consider the likelihood of an attacker discovering and exploiting this vulnerability – perhaps high, given the known issues with the software – and the potential impact, such as data loss, downtime, or reputational damage.

Post the evaluation of threats, risk assessment involves risk prioritisation. This stage is crucial as it informs the order in which risks should be addressed, helping an organisation allocate its resources effectively. In our server software example, if the server hosts critical data and the impact of an attack would be significant, this risk might be prioritised over others with lower potential impacts.

Risk assessment is not a one-off event, but rather an ongoing activity. The digital landscape is continuously evolving, with new threats and vulnerabilities emerging all the time. Therefore, risk assessments should be reviewed and updated regularly to ensure they reflect the current threat environment.

Once risks have been identified, evaluated, and prioritised, the results of the risk assessment are typically documented in a risk assessment report. This report serves as a vital tool for communication within the organisation, ensuring that everyone understands the cybersecurity risks and the actions needed to mitigate them. It might include information about the vulnerabilities identified, the possible impacts of these vulnerabilities being exploited, and recommendations for mitigating each risk.

Effectively, risk assessment serves as the compass in the turbulent journey of maintaining cybersecurity. It doesn't promise smooth sailing, but it provides the necessary direction and prepares the crew to deal with any storms that may come their way. It is the guide that informs where to build walls, where to station guards, and where to lay traps in the unending quest to protect the kingdom that is an organisation's digital infrastructure.

3.4.2.1. Quantitative and Qualitative Risk Assessments

Assessing risks in cybersecurity involves analysing complex systems and interpreting uncertain, sometimes incomplete data. Two central methodologies facilitate this process: quantitative and qualitative risk assessments. Each provides a distinct lens through which risks can be examined, and both can deliver complementary insights when applied together.

Quantitative risk assessment leans heavily on numeric data and statistical methods. This process involves assigning concrete values to the likelihood and impact of risks, often based on historical data or detailed projections. By quantifying risks, organisations can communicate about threats in clear, numeric terms, enabling data-driven decision-making.

For example, an organisation might quantify the risk of a data breach by evaluating the number of attempted breaches over a specific period and the average cost of each breach. This quantitative data can be used to calculate the potential financial loss from such breaches, helping to inform budgeting for cybersecurity measures.

On the flip side, qualitative risk assessment relies on descriptive or categorical data. This methodology focuses less on numbers and more on interpretation and judgment, providing a more holistic, nuanced view of the risk landscape. It typically involves categorising risks into groups such as 'low', 'medium', or 'high' based on expert opinion, industry benchmarks, or regulatory frameworks.

For instance, an organisation may lack detailed data about a new type of phishing attack. Instead, experts might use their knowledge to evaluate the potential impact and likelihood qualitatively, considering factors such as the sophistication of the attack, the sensitivity of the data targeted, and the organisation's current phishing defences.

Both methodologies have their place in a well-rounded risk assessment strategy. Quantitative risk assessments can provide a clear, objective basis for decision-making, helping to justify investments in cybersecurity. At the same time, qualitative risk assessments can help to capture and communicate the complexity and uncertainty inherent in cybersecurity, and they can be especially useful when hard data is lacking or unreliable.

Moreover, combining these approaches can offer a more comprehensive picture of an organisation's risk landscape. Qualitative assessments can help identify potential threats and vulnerabilities, which can then be quantitatively assessed to provide a more precise understanding of their potential impact.

Essentially, the choice between quantitative and qualitative risk assessment is not an either/or decision. Instead, organisations should seek to leverage both methodologies as part of a comprehensive, nuanced approach to understanding and managing cybersecurity risks.

3.4.2.2. Risk Matrix and Risk Heat Maps

The risk matrix, at its core, is a visual representation of uncertainty, a cartography of probable threats. Picture a grid where one axis represents the likelihood of a threat, and the other axis indicates the potential impact. The points inside the grid—a cluster of risks—are tagged with qualitative labels ranging from 'low' to 'high'. This labelling breathes life into the abstract concept of risk, enabling teams to identify, discuss and prioritise threats.

Consider an entity in the healthcare sector, wrestling with the potential threat of a data breach. By determining the probability of such an event and its potential impact—say, loss of sensitive patient information—the organisation can place this risk on the matrix. An eye-catching placement in the 'high' likelihood and 'high' impact quadrant prompts immediate action plans, driving home the urgency and the stakes involved.

The risk heat map is an evolved version of the risk matrix, bringing in the power of color to convey risk levels. Each cell on the grid dons a color—often a spectrum from cool to warm colours—corresponding to its risk level. This added visual dimension refines the understanding and communication of risks, enabling teams to instantly identify the most pressing threats.

To illustrate, imagine a software-as-a-service provider wrestling with a multitude of threats: unauthorised access, service disruptions, software vulnerabilities, to name a few. With each risk plotted on the heat map, the intensity of colours act as a beacon for risks demanding immediate attention. This simplification accelerates decision-making, guiding the allocation of resources where they matter most.

In the grand scheme of cybersecurity, the risk matrix and risk heat maps may seem like simplistic tools. However, they serve as invaluable aids for navigating the volatile landscape of threats, bridging the gap between complex data and its human understanding. With these tools, organisations can break down the abstract notion of risk into concrete, actionable components. Their adoption contributes to a culture of proactive risk management, where threats are understood, communicated and mitigated effectively, bringing resilience to the heart of operations.

3.4.3. Risk Mitigation Strategies

Risk mitigation forms the crux of an effective cybersecurity strategy. It involves making informed choices to lessen the potential harm that could arise from identified risks. It is a process that brings together the understanding of risk with actions that tackle these risks head-on. It's about navigating the labyrinth of threats, not with an armour of naivety, but with the shield of strategy and sword of action.

Diverse as the threats are, so too are the risk mitigation strategies. The choice of strategy often depends on the nature of the risk, its potential impact, and the resources at hand. Four major strategies generally guide the path towards risk mitigation: risk acceptance, risk avoidance, risk reduction, and risk transfer.

Risk acceptance is chosen when the cost of mitigation exceeds the potential loss from the risk. This is not a defeatist approach but rather a calculated decision where the likelihood and impact of the risk are deemed manageable. However, accepting a risk does not mean ignoring it. It still requires continuous monitoring to ensure that it doesn't evolve into a greater threat.

Risk avoidance, on the other end of the spectrum, involves not participating in activities that could trigger the risk. This often translates to not implementing a certain technology or not engaging in a specific business operation that is likely to invite a particular cybersecurity threat. For instance, if a software application has multiple known vulnerabilities, an organisation may choose not to use it at all.

Risk reduction is arguably the most proactive strategy, focusing on lowering the probability of risk occurrence or minimising its potential impact. This often involves implementing safeguards and controls. A common example of risk reduction is enforcing strong password policies or implementing multi-factor authentication in an organisation to reduce the risk of unauthorised access.

Lastly, risk transfer involves shifting the risk responsibility to another entity. This often takes the form of insurance policies or outsourcing certain operations to third-party vendors. For instance, an organisation may decide to transfer the risk of data storage by using a third-party cloud service provider.

Risk mitigation is not about eradicating all risks but managing them to an acceptable level. It's a balancing act that involves understanding the threat landscape, prioritising risks based on their potential impact, and deploying resources effectively. It's also an ongoing process, constantly evolving in response to new threats and vulnerabilities. As the digital world evolves, so too must our strategies for mitigating its risks. This journey demands vigilance, agility, and an unyielding commitment to protecting our cyber realms. After all, in cybersecurity, the strongest armour is preparedness, and the best offense is a good defence.

3.4.3.1. Risk Acceptance, Avoidance, Mitigation, and Transfer

Risk acceptance, avoidance, mitigation, and transfer form the cornerstone of risk management strategies in the realm of cybersecurity. A masterful blend of these strategies, tailored to an organisation's unique threat landscape, creates a shield that guards against the menacing storms of cyber threats.

In the realm of cybersecurity, risk acceptance isn't about turning a blind eye but rather making an informed decision to retain risk. Businesses may accept risk when they deem the cost of mitigation to be higher than the potential damage. However, acceptance doesn't imply ignorance. Businesses must keep track of such risks, ensuring they don't spiral out of control. For instance, a business might accept the risk of a low-level intrusion if they believe their data isn't of high value to hackers and the cost to prevent such intrusion is high.

Risk avoidance is the polar opposite of acceptance. Here, the idea is to completely remove the chance of the risk eventuating. This strategy involves not undertaking activities that could introduce the risk. For example, a business might avoid using a software that is known to have exploitable vulnerabilities that could lead to a data breach. Though this approach might seem overly cautious, it can save an organisation from a major fallout.

Risk mitigation centres on decreasing the impact or likelihood of a risk. It's about establishing controls and safeguards that diminish the threat. For instance, deploying intrusion detection systems, enforcing multi-factor authentication, and regularly updating software are strategies that fall under risk mitigation. While it might not completely eliminate the risk, it keeps it under control and manageable.

Risk transfer entails passing the risk responsibility onto another party. This can involve obtaining cybersecurity insurance or outsourcing certain operations to vendors. For instance, a company might choose to offload its data storage risks by employing a third-party cloud service provider.

Each of these strategies holds its value and is not mutually exclusive. They can be woven together into a resilient tapestry that safeguards an organisation. For example, a business might decide to accept a low-level risk, avoid a risk that threatens their core operations, mitigate a moderate risk through security enhancements, and transfer another risk through an insurance policy.

Ultimately, effective risk management is about choosing the right strategy or combination of strategies for the right risk. It's not about the total elimination of risk – an impossible feat in the ever-changing cybersecurity landscape. Instead, it's about continuously managing risks to an acceptable level while ensuring that the organisation can still operate and innovate in the digital era. In the end, the goal is to weave a protective web where every risk, whether accepted, avoided, mitigated, or transferred, finds its rightful place.

3.4.3.2. Developing a Risk Mitigation Plan

Creating a risk mitigation plan is much like navigating a ship through a storm. One must be keenly aware of the dangers, possess the tools to withstand the tumult, and have the foresight to chart the best course. A comprehensive risk mitigation plan serves as this navigational guide in the rough seas of cybersecurity.

A risk mitigation plan begins by grounding itself in the findings of the risk assessment phase. The identified risks, their possible impacts, and the vulnerabilities that could be exploited provide the foundation for devising strategies to tackle them.

The next step in building a risk mitigation plan is to prioritise the identified risks. A well-crafted plan understands that not all risks are equal – some require immediate attention while others can be addressed over time. Here, a risk's potential impact and the likelihood of its occurrence play a crucial role in determining its priority. Risks that could cause significant disruption and have a high chance of occurrence are treated with urgency.

Once risks are prioritised, appropriate mitigation strategies are assigned to each risk. These strategies could include risk acceptance, avoidance, mitigation, or transfer. The choice of strategy is guided by several factors, including the nature of the risk, the cost of the strategy, the organisation's risk tolerance, and the potential impact on business operations.

For example, a risk deemed too costly to mitigate might be transferred through a cybersecurity insurance policy. On the other hand, a risk threatening the core business operations might warrant an avoidance strategy, such as not implementing a certain technology or process.

The development of a risk mitigation plan also involves assigning responsibilities. Clearly delineated roles and responsibilities ensure accountability and foster a strong risk management culture. Specific individuals or teams should be assigned to manage, monitor, and report on each risk.

An effective risk mitigation plan is never static. It evolves, adapts, and improves. Therefore, it's important to have a review process in place. Regular reviews ensure the plan remains relevant and effective in the face of changing threats, technologies, and business operations. Review findings can lead to adjustments in risk priorities, mitigation strategies, and assigned responsibilities.

Moreover, the implementation of the risk mitigation plan should be monitored to measure its effectiveness. Key performance indicators and metrics can provide valuable insights into whether the strategies are working as intended. This feedback loop can lead to plan adjustments, enhancing the organisation's resilience against cyber threats.

At its core, a risk mitigation plan is a dynamic tool, not a one-time checklist. It must be ingrained into the very DNA of an organisation, driving its security practices, influencing its technology choices, and shaping its culture. It's not about fighting off every wave in the storm, but rather about charting the best course through it.

CHAPTER 4. CYBERSECURITY TECHNOLOGIES AND TOOLS

This chapter delves into this high-tech landscape where sophisticated solutions battle constantly evolving threats.

Cybersecurity technologies and tools range from basic firewalls to advanced artificial intelligence systems. They safeguard various aspects of digital infrastructure, from networks to applications, data to endpoints. These technologies don't just build walls around valuable assets but also watch for suspicious activity, alert on possible breaches, respond to confirmed threats, and even learn from past attacks to bolster future defences.

In essence, cybersecurity technologies and tools are the shields and swords in the struggle against cyber threats. They are the sentinels that never sleep, the detectives that never stop investigating, and the warriors that never back down.

Yet, these tools are not a panacea. They must be complemented by well-designed processes and a skilled workforce. Furthermore, their effectiveness can be influenced by factors like correct implementation, regular updates, and customisation according to specific business contexts.

As we explore the diverse universe of cybersecurity technologies and tools, we will uncover their functions, learn their mechanics, appreciate their strengths, and acknowledge their limitations. But most importantly, we will understand how they weave into the broader tapestry of cybersecurity strategy. The journey awaits. Let's begin.

4.1. Firewalls and Intrusion Prevention Systems

Firewalls and Intrusion Prevention Systems (IPS) are foundational tools in the cybersecurity toolbox. Often viewed as the first line of defence, these technologies form the frontline guardians of the cyber realm, acting as gatekeepers between a network and the potential dangers lurking in the wider internet.

A firewall, one of the oldest types of cybersecurity tools, essentially works as a network traffic controller. It scrutinises incoming and outgoing traffic based on predefined security rules. This ruleset can range from simple to complex, and the firewall enforces these rules like a steadfast guard, allowing or denying traffic based on its adherence to the established regulations. Firewalls come in different forms: network firewalls that guard the entire network, host-based firewalls that protect individual machines, and cloud firewalls that secure cloud-based resources.

While firewalls function primarily based on pre-defined rules, Intrusion Prevention Systems (IPS) add another layer of sophistication to this security mechanism. An IPS doesn't just evaluate network traffic against established rules; it also analyses the patterns and behaviour of the traffic. The underlying purpose is to detect and prevent intrusion attempts. Intrusion Prevention Systems can identify signs of an attack even when such activity doesn't explicitly break any rules, providing an extra layer of protection.

For instance, consider a scenario where an external server starts making rapid-fire connection requests to various computers in a network. A traditional firewall might allow this activity if its rules don't specifically forbid it. However, an IPS would recognise this behaviour as a potential network scanning activity – a common prelude to a more serious attack – and can block the suspicious traffic in real time.

Yet, as powerful as firewalls and IPSs are, they are not fool proof. Sophisticated attackers can sometimes find ways to bypass these defences, especially if the systems are not properly configured or updated. Therefore, it's crucial for organisations to keep these tools up-to-date and fine-tune their configurations based on evolving threat landscapes.

Moreover, firewalls and IPSs are only a part of the overall security ecosystem. They need to be complemented by other tools such as antivirus software, data encryption, secure access controls, and continuous monitoring systems to achieve comprehensive cybersecurity.

Firewalls and Intrusion Prevention Systems thus provide the critical, foundational layer of cybersecurity. They exemplify how cybersecurity isn't just about building walls but also about patrolling borders, detecting intrusions, and responding proactively. In the next sections, we will explore other technologies and tools that fortify this foundation and ensure a 360-degree cybersecurity strategy.

4.1.1. Understanding Firewalls: Types and Functions

A firewall, an aptly named concept inspired by a physical barrier designed to halt the spread of fire, is a network security system that monitors and controls network traffic based on an organisation's previously established security policies. It serves as a barrier between a trusted internal network and untrusted external networks, such as the internet.

Firewalls come in various forms, each with its strengths, weaknesses, and ideal use cases. Broadly, firewalls can be classified into four types: packet-filtering firewalls, stateful inspection firewalls, proxy firewalls, and next-generation firewalls.

Packet-filtering firewalls, the most traditional type, operate at the network level of the OSI model. They analyse packets (the basic units of data transfer) and allow or block them based on rules relating to IP addresses, protocol, or port number. While they are fast and efficient, they lack the sophistication to detect complex threats.

Stateful inspection firewalls, on the other hand, monitor the state of active connections and use this information to determine which network packets to allow. They offer more security than packet-filtering firewalls because they track ongoing communication sessions, making it more difficult for an unauthorised user to sneak past them.

Proxy firewalls, often referred to as application-level gateways, act as intermediaries for requests from one network to another. They filter incoming traffic between the network and the traffic source, providing a high level of security. Proxy firewalls can inspect the entire application data portion of a packet, enabling them to block specific websites, recognise malicious software, or prevent attempts to exploit application-layer vulnerabilities.

Next-Generation Firewalls (NGFWs) are the most advanced type. They incorporate the functions of the previous types, but with more capabilities such as deep packet inspection, intrusion prevention systems, and application awareness. These firewalls can identify and block complex attacks by enforcing security policies at the application, port, and protocol levels. They are often used in enterprise settings where robust network security is a necessity.

While firewalls come in different types, they all serve a common function - to control network traffic. They do so through a set of defined rules, also known as access control lists (ACLs). These rules specify which traffic should be allowed or denied based on factors like IP addresses, port numbers, and the protocol used. For example, a rule might allow all incoming emails while blocking traffic from a specific IP address known to be a source of malicious activity.

Firewalls can also provide additional functionalities. Some firewalls include Virtual Private Network (VPN) support, enabling secure remote access to a network. Others might offer user authentication, adding another layer of access control.

In managing these various functions, a key part of the firewall's role is to maintain a log of its activity. These logs, detailing the traffic passing through the firewall and any detected threats, can provide valuable data for security audits and incident investigations.

Despite the robust capabilities of firewalls, no single technology can guarantee total security. Firewalls form a significant part of a layered defence strategy known as defence in depth. Other components, like intrusion prevention systems, secure gateways, and end-point protection, complement firewalls and contribute to an effective security posture.

In conclusion, firewalls, in their various forms, offer crucial functions in network security by monitoring and controlling network traffic. They act as vigilant gatekeepers, keeping a close eye on all network communications, halting any data flow that doesn't comply with established security rules. Understanding their functions and how to leverage them is a vital part of building and maintaining a secure network environment.

4.1.2. Intrusion Prevention Systems: Monitoring and Blocking Threats

One cannot overlook the importance of Intrusion Prevention Systems (IPS) in our age of ever-evolving cyber threats. An IPS is a powerful tool that significantly enhances a network's defences, scrutinising all incoming and outgoing traffic, flagging suspicious activity, and taking decisive actions to neutralise threats.

An IPS inspects network traffic in real-time and uses predefined rules or learned patterns to identify and halt malicious activities. While IPS shares similarities with its sibling technology, Intrusion Detection Systems (IDS), they are fundamentally distinct in one critical area: while IDS is a passive observer that alerts administrators of potential threats, an IPS takes an active role and can block threats automatically.

At the core of an IPS are its threat detection methods. These can be signature-based, anomaly-based, or behaviour-based. Signature-based detection is akin to a criminal fingerprint database. It uses known patterns of data, or 'signatures' of recognised threats, to detect malicious activity. Although extremely effective against known threats, this method struggles to identify novel attacks.

Anomaly-based detection, on the other hand, focuses on establishing a baseline of normal network behaviour. Any deviations from this 'normality' are treated as potential threats and are promptly flagged. This method can detect previously unknown threats, making it valuable in a landscape of ever-evolving cyber-attacks.

Behaviour-based detection further builds on anomaly detection. It not only detects deviations from a baseline but also learns and adapts over time, reducing the likelihood of false positives and increasing its accuracy in identifying threats.

118

An essential feature of an IPS is its ability to respond to detected threats swiftly. This capability to take immediate action, such as dropping malicious packets, resetting a connection, or blocking traffic from suspect IP addresses, sets IPS apart from IDS. For a cybersecurity team, the ability of an IPS to neutralise threats can drastically reduce response time and mitigate potential damage.

IPS devices can be network-based, wireless, network behaviour analysis (NBA), or host-based. Network-based IPS (NIPS) monitors the entire network for suspicious traffic. Wireless IPS (WIPS) focus on wireless networks, detecting rogue access points and unauthorised logins. NBA systems examine network traffic to identify threats that generate unusual traffic flows, such as Distributed Denial of Service (DDoS) attacks. Host-based IPS (HIPS) operates on a single host, monitoring it for suspicious activity.

Although powerful, an IPS is not a standalone solution. Just like a firewall, it forms part of a broader security strategy. It should be complemented with other security measures such as firewalls, security information and event management (SIEM) systems, and end-point protection to create a comprehensive defence against a vast range of cyber threats.

Through their real-time inspection of network traffic and active blocking capabilities, Intrusion Prevention Systems serve as an invaluable component in cybersecurity. As cyber threats continue to increase in sophistication, the role of IPS in detecting and neutralising these threats becomes ever more crucial. By understanding the functions of an IPS, organisations can effectively implement them into their security architecture, bolstering their defences, and enhancing their resilience against cyber threats.

4.2. Antivirus and Antimalware Solutions

Entering the world of cybersecurity technologies, one encounters a multitude of tools designed to shield digital realms from a spectrum of threats. Among the most vital are antivirus and antimalware solutions. These quintessential defences have evolved over time, keeping pace with the cunning creativity of threat actors. Their purpose remains simple yet profound: protect systems by detecting, quarantining, and eliminating malicious software, ensuring the safety and integrity of our digital ecosystems.

Antivirus solutions, part of every cybersecurity toolkit since the dawn of the internet age, primarily concentrate on guarding against computer viruses. However, the term 'antivirus' is often used interchangeably with 'antimalware' as many antivirus solutions have broadened their range to counter a wider set of malicious software, including worms, trojans, ransomware, and spyware.

Functionally, antivirus and antimalware tools are deployed to perform real-time scanning, ensuring each file is checked as it is accessed or modified. This proactive approach is often called 'on-access scanning' or 'resident scanning.' The goal is to stop malicious software in its tracks before it can execute and compromise the system.

In addition to real-time defence, scheduled scans are performed to check every file in the system. This 'full system scan' is a comprehensive sweep, searching for any hidden or dormant threats that may have slipped past real-time defences. It is analogous to a thorough medical check-up, providing an exhaustive health report of the system.

The backbone of these tools is their detection methods, predominantly signature-based and behaviour-based detection. Signature-based detection uses a database of known malware signatures – unique strings of data within the code of each malicious program. When a file matches a known signature, it's flagged as a threat. This method is powerful against known threats but struggles with new, unknown ones.

Behaviour-based detection, or heuristic analysis, tackles this limitation. It monitors the behaviour of programs, looking for actions typical of malicious software. For instance, a program attempting to modify system files or reading from sensitive areas in memory might be flagged. This method provides protection against zero-day threats – new malware that hasn't yet been analysed and added to signature databases.

Sandboxing is another feature found in advanced antivirus and antimalware solutions. It allows potentially suspicious programs to execute in a safe, isolated environment, preventing them from affecting the rest of the system. This feature enables the software to observe the behaviour of the program without risking the security of the system.

While these technologies provide robust protection, they must be paired with other security measures to ensure comprehensive coverage. Firewalls, Intrusion Prevention Systems, encryption, and safe browsing practices are essential complements to antivirus and antimalware solutions.

With the proliferation of digital devices and the ever-growing reliance on digital data, the importance of antivirus and antimalware solutions cannot be overstated. Their role in safeguarding systems from malicious threats remains crucial. As cybersecurity professionals, understanding these technologies enables us to better protect our digital fortresses, ensuring the safety, privacy, and trust that our digital world hinges upon.

4.2.1. How Antivirus and Antimalware Work

Engaging with the process of defending digital systems from threats, the inner workings of antivirus and antimalware solutions may seem akin to an intricate dance. Each step is orchestrated to detect, prevent, and eliminate the threats that incessantly attempt to disrupt the rhythm of our digital lives.

At their core, antivirus and antimalware solutions operate on principles of identification and remediation, propelled by a set of detection methods. The first and perhaps the most traditional method, signature-based detection, operates much like a detective seeking known culprits. In this case, the culprits are malicious programs, each bearing unique lines of code known as 'signatures.'

Antivirus programs maintain a database of known malware signatures, gleaned from the examination of millions of malicious software samples. When the antivirus software scans a file, it calculates the file's signature and checks if it matches any in its database. A match indicates the presence of a known threat. However, this method has its limitation as it can only detect threats that have been previously identified and included in the database.

To tackle the challenge of new and unknown threats, behaviour-based detection, also known as heuristic-based detection, is employed. Rather than solely relying on known signatures, this method monitors programs and files for suspicious behaviour. Actions such as unauthorised attempts to access system files, abnormal resource usage, or the modification of other programs could all raise red flags. By recognising suspicious behaviour, antivirus and antimalware solutions can identify and stop unknown threats, even if they've never been seen before.

Both these methods are used in conjunction for real-time protection, continuously scanning files when they're accessed or modified. This shields the system proactively, aiming to stop malware before it has a chance to cause damage. Scheduled, full-system scans are also performed periodically to catch any threats that may have slipped past real-time scanning.

Many contemporary antivirus and antimalware solutions also feature sandboxing. This method allows potentially harmful programs to run in a controlled, isolated space, separate from the system's core functions. It's a safety measure to study the behaviour of unknown files without risking system integrity.

Another critical function of antivirus and antimalware solutions is remediation, or the process of dealing with detected threats. Once a threat is identified, the software aims to remove it and clean any infected files. In cases where a file cannot be cleaned, the software typically quarantines it, isolating it from the rest of the system to prevent further harm.

As threats evolve, so too must the mechanisms that counter them. Despite their sophistication, antivirus and antimalware tools should not be seen as a silver bullet. They form a critical layer of defence in a comprehensive, multi-layered security strategy. As users and defenders of digital systems, understanding their workings equips us to better navigate the ever-changing cybersecurity landscape. The dance continues, and with each step, we strive to remain one beat ahead.

4.2.2. Choosing the Right Antivirus Solution

Delving into the world of antivirus and antimalware solutions, it's evident that there's no one size fits-all approach. Choosing the right solution requires a combination of understanding the unique needs of your digital landscape, being aware of the strengths and weaknesses of different software, and aligning your choice with your risk tolerance and digital habits.

When considering an antivirus solution, one of the initial steps is to analyse your needs. Are you an individual who only needs basic protection for a single device, or are you responsible for the digital security of a large organisation with a vast network of interconnected systems? The scale and complexity of your digital environment can significantly influence your choice.

The operating system you're using can also play a pivotal role in your decision. Not all antivirus solutions are compatible with all operating systems. Therefore, it's essential to ensure the software you select is compatible with and optimised for your operating system.

Another vital factor is the comprehensiveness of the solution. Basic antivirus software typically focuses on traditional threats like viruses, worms, and Trojans. However, as digital threats continue to evolve, there's an increasing need for a more comprehensive solution that can tackle a wider array of threats, including ransomware, spyware, and adware. A solution offering real-time protection, behaviour-based detection, and remediation capabilities should be given significant consideration.

Performance impact is another crucial consideration. Some antivirus programs can be resource-intensive, slowing down your system while they scan. While comprehensive protection is vital, it shouldn't come at the expense of system performance. Reviews and benchmarks can provide insights into the performance impact of different solutions.

Furthermore, it's worthwhile to consider the solution's user interface and ease of use. A cluttered or overly complicated interface can make using the software a frustrating experience and deter you from fully utilising its features. You'll want a solution that's straightforward, easy to navigate, and offers comprehensive customer support in case of any issues or inquiries.

While free antivirus solutions may seem tempting, it's essential to recognise that they often lack the comprehensive protection and support that comes with paid options. Free solutions can be a viable option for basic protection, but for complete peace of mind, a premium solution is typically more reliable.

Lastly, before making a decision, it can be beneficial to take advantage of free trials. These allow you to evaluate how well a solution fits your needs, its impact on your system, and the effectiveness of its protection without making an immediate financial commitment.

In this vast sea of cybersecurity tools, choosing the right antivirus solution can feel like an overwhelming endeavour. Yet, it is a necessary step in ensuring a resilient digital life. It's about more than just picking software; it's about crafting a personalised shield, robust enough to defend against the storms of cyber threats, and yet, flexible enough to allow you to navigate the digital world freely and confidently.

4.3. Encryption and Public Key Infrastructure

If we view the digital world as an open sea, teeming with threats lurking below the surface, then encryption is akin to the sturdy, impenetrable hull of a ship, warding off the hostile waters. Combined with Public Key Infrastructure (PKI), this duo forms the backbone of digital privacy and secure communications in an ever-more-connected cyber landscape.

Encryption, at its core, is a process that transforms readable data (plaintext) into an unreadable format (ciphertext) using an algorithm and an encryption key. This assures the confidentiality and integrity of data as it moves across networks or rests on storage devices, making it inaccessible to anyone who does not possess the corresponding decryption key. It's a digital padlock that keeps prying eyes away from your data, ensuring that even if it falls into the wrong hands, it remains unintelligible.

On the other hand, PKI is a framework for services that enable secure, scalable, and manageable distribution and identification of public encryption keys. It uses a pair of keys— a public key for encrypting or verifying signatures, and a private key for decrypting or signing. PKI brings trust to digital transactions by verifying the identities of parties involved and creating a secure channel for data transmission.

Let's delve into various aspects of encryption and PKI.

Firstly, there are two main types of encryption: symmetric and asymmetric. Symmetric encryption uses the same key for both encryption and decryption, making it faster but less secure as anyone with the key can decrypt the data. Asymmetric encryption uses a pair of mathematically linked keys, one for encryption (public key) and one for decryption (private key), making it more secure but slower. PKI typically utilises asymmetric encryption.

PKI provides a critical service in the digital world: digital certificates. These electronic "passports" verify that a public key belongs to the individual or entity stated in the certificate. These certificates, issued by a trusted third party known as a Certificate Authority (CA), help to mitigate the risk of Man-in-the-Middle attacks, where an attacker intercepts and potentially alters communication between two parties.

But the strength of PKI isn't just in the encryption; it's also in the intricate web of trust it establishes. A robust PKI ensures that all users are who they claim to be, that all websites are legitimate, and that all data transfers are secured. This framework doesn't just protect data; it builds a trustworthy and secure digital environment.

In practice, encryption and PKI are ubiquitous. From safeguarding our online purchases to protecting the confidentiality of our emails, these technologies are embedded in our daily digital interactions. They are critical components of technologies such as Secure Sockets Layer (SSL) and Transport Layer Security (TLS) protocols, which are foundational to secure web browsing and other secure communications.

Despite their robustness, it's essential to understand that encryption and PKI are not magic bullets. They need to be complemented with other security measures and policies to provide effective cybersecurity. Additionally, they require proper management and governance. For example, private keys need to be securely stored and handled to prevent unauthorised access, and digital certificates need to be regularly renewed and revoked when no longer needed.

In the vast sea of cybersecurity, encryption and PKI serve as our sturdy vessel, protecting us from the relentless waves of threats. They stand as silent guardians, ensuring that our digital journey is not only secure, but also trustworthy. The future of cybersecurity will undoubtedly bring new challenges and risks, but with the steadfast shield of encryption and the validating power of PKI, we can navigate the tumultuous waters with confidence.

4.3.1. Understanding Encryption: Symmetric and Asymmetric

Encryption, the transformative act of turning readable data into an unreadable format, forms a linchpin in cybersecurity efforts around the globe. By rendering data unintelligible to unauthorised parties, encryption ensures the confidentiality and integrity of information as it voyages through digital spaces. It's an elegant yet potent methodology, with its two primary forms, symmetric and asymmetric encryption, representing the dual facets of this fascinating approach to securing data.

Symmetric encryption, the simpler and older of the two types, operates using a single key. This key, like a special cipher, both locks (encrypts) and unlocks (decrypts) the data. Imagine a literal, physical box that you wish to secure. With symmetric encryption, the same key that locks the box also unlocks it. It's an efficient and speedy method, but it carries with it an inherent risk: if the key falls into the wrong hands, they can both lock and unlock the box at will.

Common symmetric encryption algorithms include Advanced Encryption Standard (AES), Data Encryption Standard (DES), and RC4. AES, in particular, is widely regarded for its superior security and is used by the U.S. government for encrypting classified information.

Contrastingly, asymmetric encryption uses two distinct yet related keys, adding an extra layer of security. In this case, one key is used to lock (encrypt) the box, and another, separate key is used to unlock it (decrypt). This technique is also known as public-key cryptography. The beauty of asymmetric encryption is that you can freely distribute the public key (used for encryption) without endangering the secrecy of the private key (used for decryption).

Rivest-Shamir-Adleman (RSA) and Elliptic Curve Cryptography (ECC) are examples of asymmetric encryption algorithms. RSA, which is widely used in web browsers and email, is lauded for its security, but it requires more computational power, making it slower than its symmetric counterparts.

Asymmetric encryption brings an additional boon to the cybersecurity realm: digital signatures. By using the private key to encrypt, instead of decrypt, an individual can apply a unique digital signature to data. This digital signature, which can be publicly verified using the associated public key, provides proof of the data's authenticity and integrity, lending credibility and accountability to digital interactions.

However, asymmetric encryption also has its challenges. Managing the public and private keys demands additional caution and security measures. It is also computationally intensive, which can be a disadvantage in resource-limited environments or when dealing with large amounts of data.

Together, symmetric and asymmetric encryption form a potent toolkit for cybersecurity. They protect data confidentiality, uphold integrity, and in the case of asymmetric encryption, provide non-repudiation through digital signatures. Grasping these fundamental concepts illuminates the landscape of data protection and provides the groundwork for navigating the complexities of cybersecurity in an increasingly connected world. They represent the double-edged sword wielded in the eternal vigilance against cyber threats, each playing its part in the grand strategy of data defence.

4.3.2. Public Key Infrastructure: Roles and Elements

Navigating the complex domain of cybersecurity entails grasping a broad spectrum of concepts and tools. Among these, Public Key Infrastructure (PKI) represents an indispensable framework to uphold security within digital communications. Rooted in the principles of asymmetric encryption, PKI enables the secure exchange of data over unsecured and widely accessible networks like the internet.

At its core, PKI consists of roles, policies, hardware, software, and procedures working in unison to enable the creation, distribution, management, storage, and revocation of digital certificates. These digital certificates, similar to digital passports, offer a way to prove and uphold the identities of individuals, devices, and websites in the virtual world.

A critical element of PKI is the Certificate Authority (CA), a trusted third party that issues and manages digital certificates. Acting as a kind of digital notary, the CA validates entities' identities, issues certificates attesting to that identity, and maintains a directory of the certificates issued. If a certificate becomes compromised or the holder's identity changes, the CA is also responsible for revoking the certificate.

Another vital component is the Registration Authority (RA). It serves as the intermediary between users and the CA, verifying the identities of entities before they are issued a digital certificate by the CA. The RA takes on the duty of performing identity checks, reducing the workload of the CA and adding an extra layer of security.

To balance security and performance, a PKI system employs both asymmetric and symmetric encryption. When initiating communication,

130

asymmetric encryption is first used to securely exchange a symmetric encryption key. After the exchange, the faster symmetric encryption takes over for the rest of the communication. This two-step process ensures robust security while optimising computational resources.

Other integral parts of PKI include the certificate database, which stores and manages all digital certificates and their statuses, and the certificate store, where an individual or an entity keeps their own digital certificates and private keys.

The role of PKI in the digital sphere is profound. It underpins secure electronic transactions, protects the privacy of data transmitted over networks, ensures data integrity, and provides non-repudiation and authentication. As an example, when you see the 'lock' icon in your web browser, it means a PKI-issued digital certificate protects your connection to that website.

Understanding the roles and elements within PKI is not just for cybersecurity professionals. As more of our lives and businesses become digital, appreciating how PKI functions equips us to better navigate and secure our activities in the digital realm. From everyday web browsing to online banking and e-commerce, PKI quietly safeguards our digital lives and gives us the confidence to engage in a world where cyber threats are an ever-present reality.

4.4. Security Information and Event Management (SIEM) Systems

In the cybersecurity landscape, establishing a watchful eye over every nook and cranny of an organisation's network is paramount. Security Information and Event Management (SIEM) systems stand as the all-seeing overseers in this domain, offering an amalgamation of capabilities that enable a comprehensive and real-time view into the security state of an enterprise's digital assets.

As its name implies, SIEM merges two crucial facets of cybersecurity: Security Information Management (SIM) and Security Event Management (SEM). Together, these elements generate a unified solution that collects and analyses log and event data in real time, identifies anomalies and threats, and enables swift response to potential security incidents.

At the heart of SIEM systems are the log data inputs from multiple sources across a network. This vast pool of information, originating from devices, applications, systems, and even cloud-based infrastructure, is captured, normalised, and aggregated. This allows SIEM to provide visibility across the entire organisational infrastructure, transforming raw data into meaningful insights.

SIEM systems rely on rules and algorithms to sift through the aggregated data. This continuous process seeks out anomalies or patterns that may indicate a cybersecurity incident. Recognised events range from multiple failed login attempts (which could signify a brute force attack) to sudden changes in network traffic (possibly suggesting a Denial of Service attack).

132

Furthermore, SIEM platforms leverage threat intelligence feeds, providing up-to-date information about known malicious IP addresses, URLs, or files. This integration empowers SIEM systems to make better decisions when flagging potential threats.

A fundamental capability of SIEM platforms is their ability to respond to identified threats. They can be configured to automatically react when certain conditions are met, such as blocking IP addresses or disabling user accounts. Alternatively, they can simply alert security personnel to take action, equipping them with the necessary information to address the threat.

The power of SIEM also extends to compliance reporting. Given their extensive data collection and analysis capabilities, these systems can produce detailed reports that help organisations demonstrate compliance with various cybersecurity regulations. For instance, a SIEM system can quickly generate a report that shows all actions taken by a specific user, facilitating internal audits or inquiries.

Having a SIEM system is akin to having a powerful, tireless sentinel that ceaselessly watches over a digital landscape. However, it's not a "set and forget" type of tool. To optimise its effectiveness, a SIEM system must be properly configured, regularly updated, and managed by trained professionals. The robustness of a SIEM solution depends on the quality of its rules and the expertise of the team working behind it.

As the complexity and volume of cyber threats escalate, having a SIEM solution becomes increasingly vital. Armed with this advanced tool, organisations can attain an enhanced level of visibility, detection, and response, significantly improving their overall cybersecurity posture.

4.4.1. The Role of SIEM in Cybersecurity

Imagine a nerve centre that sifts through a constant stream of information, distinguishing the crucial from the trivial, detecting subtleties that might otherwise go unnoticed. This is the essence of Security Information and Event Management (SIEM), a robust tool that stands as the sentry for an organisation's cybersecurity.

The role of SIEM in cybersecurity is multi-faceted and dynamic, encompassing real-time threat detection, incident response, log management, and compliance reporting. But beneath these functions, SIEM has a more fundamental purpose: it brings together a wide range of data in one centralised location, allowing for a more cohesive and comprehensive analysis of the organisation's security posture.

Akin to an observatory, a SIEM system surveys a broad horizon. It pulls together data from various sources, including firewalls, antivirus software, intrusion detection systems, and others. It even extends its reach into cloud-based infrastructure, creating an interconnected web of surveillance across the digital expanse of an organisation.

With this immense wealth of information, SIEM performs real-time analysis, parsing through the data to uncover patterns, trends, or anomalies that might indicate a security incident. This is the pulse of SIEM's operation—constantly monitoring, constantly analysing, constantly vigilant.

SIEM uses predefined rules and dynamic algorithms to detect anomalies. It scrutinises activities such as repeated failed login attempts, changes in network traffic, and actions that deviate from the usual behaviour patterns of users. And as it uncovers potential threats, it sends alerts to security teams, providing detailed information to facilitate swift, informed responses.

The role of SIEM extends into automated responses as well. Configured to react when certain conditions are met, it can take immediate actions, like blocking an IP address or disabling a user account, thus mitigating the impact of a security incident.

Another major function of SIEM lies in its log management capability. As a vast repository of log data, SIEM serves as an invaluable resource for investigations, forensic analyses, or compliance audits. It enables an organisation to revisit past events, trace activities, or substantiate compliance with security policies and regulations.

Furthermore, the compliance reporting capabilities of SIEM help organisations maintain transparency and accountability. By generating detailed reports, SIEM demonstrates adherence to various cybersecurity regulations, simplifying the process of audits and reviews.

Yet, SIEM is not an autopilot tool. Its effectiveness hinges on proper setup and consistent management. The rules that govern its operation must be thoroughly defined and regularly updated to reflect the evolving threat landscape. The accuracy of its analysis relies heavily on the quality of the data it processes and the sophistication of its configuration.

Therefore, SIEM requires a team of skilled professionals who understand the system's capabilities and limitations, and who can fine-tune its operation as needed. The true power of SIEM is unlocked when technology and expertise converge, forming a formidable barrier against cybersecurity threats.

To encapsulate, SIEM is a powerhouse in cybersecurity, a guardian that tirelessly monitors, detects, and responds. Through its multi-layered approach, it reinforces the organisation's defence, contributes to the maturity of its security posture, and ultimately, enables the organisation to navigate the challenging waters of cybersecurity with greater confidence and preparedness.

4.4.2. Deploying and Managing a SIEM Solution

The process of deploying and managing a Security Information and Event Management (SIEM) solution mirrors the art of cultivating a well-balanced ecosystem. It is a pursuit that requires precision, vigilance, and a deep understanding of the environment at hand. The efforts put into the deployment and management of a SIEM solution significantly determine its effectiveness and the overall cybersecurity posture of an organisation.

The initial step in deploying a SIEM solution is a comprehensive understanding of the organisation's environment, its assets, and its unique requirements. This stage is akin to a landscape architect surveying the land before designing a layout. The organisation must delineate its critical assets, the key data sources that need to be connected to the SIEM system, and the specific security threats that it wants to focus on.

Next, the organisation needs to establish the policies and rules that the SIEM system will follow. These policies form the blueprint for the SIEM's operations, defining what constitutes normal and abnormal behaviour within the organisation's network. The rules guide the SIEM's detection algorithms, dictating how the system should react when it spots potential security incidents.

It's essential to ensure that the SIEM system integrates smoothly with the existing infrastructure, including firewalls, intrusion detection systems, antivirus software, and more. This allows the SIEM system to pull together all the relevant security data in one place, enhancing visibility and facilitating more effective threat detection and response.

In terms of management, ongoing maintenance is paramount for the effective operation of a SIEM solution. This includes regular system updates, rules adjustments, and policy revisions to accommodate changes in the organisation's infrastructure and the evolving threat landscape.

One of the fundamental aspects of SIEM management is tuning. Tuning refers to the adjustment of the SIEM's rules and thresholds to reduce false positives and false negatives. This is an iterative process that often involves analysing past incidents, examining the details of alerts, and making necessary adjustments to improve the system's accuracy over time.

Another essential management task is conducting regular audits of the SIEM system itself. These audits help confirm that the system is functioning as expected, that it is processing and correlating data correctly, and that it is effectively identifying security incidents.

In addition, SIEM management involves reviewing and responding to the alerts that the system generates. Timely response is crucial in mitigating potential security incidents and minimising their impact. Over time, the analysis of these alerts and the corresponding responses can provide valuable insights for further tuning of the SIEM system.

Training is another essential component in managing a SIEM solution. As the system evolves and matures, the security team must stay abreast of its capabilities, learn how to interpret its alerts effectively, and understand how to respond to various incidents.

Ultimately, the deployment and management of a SIEM solution is not a one-time effort but a continuous process. It requires a proactive approach, regular review, and consistent fine-tuning. It is a dance of sorts, a dynamic interplay between the organisation and the SIEM system, striving for harmony in the complex, ever-changing realm of cybersecurity. With the right focus and commitment, a well-managed SIEM solution can significantly enhance an organisation's security posture, providing robust defence and fostering a culture of cybersecurity awareness

4.5. Multi-Factor Authentication (MFA)

Identity is power. The ability to conclusively prove one's identity forms the basis of trust in the digital world. Yet, traditional methods of authentication, such as passwords, have proven to be fallible, frequently falling victim to brute force attacks, phishing, and other hacking techniques. The quest for stronger identity verification mechanisms has given rise to Multi-Factor Authentication (MFA), a security approach that's akin to a robustly constructed fortress, built to withstand diverse types of onslaughts.

MFA stands as a citadel of security, combining multiple layers of defence to provide enhanced protection. The essence of MFA lies in its requirement for multiple forms of verification, with each form serving as an individual line of defence. These verification factors typically fall into three categories: something you know (like a password), something you have (like a physical token or a mobile device), and something you are (biometric data, such as fingerprints or facial recognition). It's as though the fortress of MFA has multiple doors, each requiring a different key to open.

The strength of MFA lies in its multi-layered approach. Should one factor be compromised, the others still stand firm, protecting the user's data. For example, even if a malicious actor manages to steal a password, they will be thwarted if they also need to bypass biometric verification or provide a unique code sent to the user's mobile device.

MFA has gained widespread adoption in various forms. For instance, banks often use MFA for online banking, requiring users to input a password and a one-time code sent via text message. Social media platforms also offer MFA options, typically in the form of a password combined with a push notification to a mobile device.

However, deploying MFA is not without its challenges. For the fortress to stand strong, each component must be well-maintained and function seamlessly. User experience is a key consideration – security measures that are too onerous can lead to frustration and resistance from users. Therefore, striking the right balance between security and usability is crucial in the design and implementation of MFA systems.

Furthermore, each form of verification needs to be secure in its own right. Biometric data, for example, needs to be stored and processed securely to prevent theft and misuse. Likewise, hardware tokens need to be physically secure, and software tokens need to be protected against hacking.

Effective management of an MFA system involves ongoing monitoring and maintenance. Regular audits can help identify any issues or vulnerabilities, and periodic user training ensures that everyone knows how to use the system correctly. As technology evolves, MFA systems may need to be updated or replaced to maintain their effectiveness.

Implementing MFA is like building a strong fortress around your digital identity. It's a commitment to robust security, recognising the crucial importance of identity in the digital world. When implemented and managed effectively, MFA can provide a strong line of defence against cyber threats, protecting both individual users and the organisation as a whole. Through its multi-layered approach, MFA empowers users to assert their digital identity confidently, navigating the interconnected landscape of the internet with a greater sense of security.

4.5.1. The Need for MFA in Today's Cyber Threat Landscape

As we journey deeper into the digital age, the monumental rise in cyber threats leaves no room for complacency. Passwords alone, once the stalwarts of digital defence, have been exposed as porous shields, too often easily penetrated by the ceaseless barrage of cyber-attacks. Hence, the call to action is clear – we must bolster our fortifications. Multi-Factor

Authentication (MFA), akin to a multi-layered shield, stands tall in the face of these challenges, offering a more formidable line of defence.

The crux of the problem lies in the inherent vulnerabilities of password-based authentication. They are often easily guessed, predictably patterned, or unwittingly shared, rendering them ineffective against the brute force of modern hacking techniques. Moreover, in the bustling digital marketplace of illicit activity, stolen passwords are commodities that are bought, sold, and traded. In this digital chess game, it's clear that the password is the pawn – easily sacrificed and offering little protection to the king – your data.

MFA steps into this void, providing much-needed reinforcement to the front lines of cyber defence. With its emphasis on multiple verification layers, MFA can thwart even determined attackers. If one layer of defence is breached, another stands ready. This resilience, coupled with the diversity of the verification factors used, creates a formidable challenge for cybercriminals. It's as though MFA has taken the chessboard and added layers, turning a two-dimensional game into a three-dimensional challenge.

Furthermore, the rise of remote work and the Internet of Things (IoT) has dramatically expanded the battlefield. As the boundaries of networks dissolve, securing each access point becomes more critical, and more challenging. MFA presents an effective solution, providing enhanced security regardless of location or device.

Yet, the changing face of cyber threats requires vigilance and evolution. The shifting strategies of cybercriminals necessitate a defence that can adapt and respond. Just as castle walls were eventually outmatched by the development of artillery, so too must MFA continue to evolve in response to the changing tactics of attackers.

MFA must stand ready to face emerging threats. Biometric data, although an effective authentication factor, can be compromised and then remains forever vulnerable – you can change a password, but you can't change your fingerprints. Furthermore, as advanced technology such as deepfakes continues to develop, even biometric verification methods may be challenged.

While it's clear that MFA offers a significant enhancement to digital defence strategies, it's equally clear that we cannot rest on our laurels. Continued innovation and vigilance are essential in maintaining the efficacy of MFA. The chess game continues, but with MFA, we're better equipped for the match. It is an essential tool in our arsenal, a dynamic shield in the face of ever-evolving cyber threats.

4.5.2. Implementing MFA: Factors to Consider

The implementation of Multi-Factor Authentication (MFA) isn't as straightforward as turning on a switch; it's more akin to navigating a labyrinth. Multiple paths present themselves, each leading to different outcomes. To navigate effectively, a variety of factors must be weighed and considered.

For starters, an organisation needs to assess the nature and sensitivity of the data that it handles. Medical records, financial transactions, or classified information necessitate a more robust approach to authentication. A company dealing in such data must strive for an MFA solution that incorporates advanced methods such as biometrics and hardware tokens.

In contrast, an entity that deals with less sensitive information could opt for simpler, less intrusive forms of MFA, such as SMS codes or application-based one-time passwords (OTP). The sensitivity and value of the information being protected are paramount in determining the level of security required.

Convenience, user experience, and adoption rates are other crucial factors to consider. MFA is only effective if it is consistently used. If an MFA solution is overly complicated or burdensome, users may seek workarounds, or worse, avoid using the system altogether. Striking a balance between usability and security is an art form that requires careful consideration.

Another important aspect is the cost associated with the deployment and maintenance of MFA. From licensing fees to the infrastructure required for managing hardware tokens, costs can add up quickly. Not all MFA solutions are created equal, and the most expensive option is not always the most effective. Therefore, organisations need to meticulously analyse their budgets and choose a solution that delivers the best value.

Cybersecurity preparedness must also be factored in when considering an MFA implementation. The organisation's IT staff must have the expertise to set up, monitor, and maintain an MFA system. In addition, employees need to be trained on how to use the system effectively and safely.

Finally, the ability of an MFA system to integrate with the organisation's existing infrastructure is critical. This could involve interoperability with the existing hardware and software, or compatibility with the cloud and other digital platforms. Seamless integration not only simplifies the deployment process but also reduces potential disruption to the business operations.

Implementing MFA is not a one-size-fits-all endeavour. The journey is unique for each organisation, dictated by its individual needs and circumstances. It's a strategic game of balancing security, convenience, costs, and compatibility. By making informed decisions and adapting as necessary, organisations can navigate the labyrinth successfully and fortify their defences against the relentless tide of cyber threats.

4.6. Endpoint Detection and Response (EDR)

It is no longer sufficient to merely protect a network's perimeter. Like an immune system not only preventing but also detecting and responding to infections, a comprehensive security strategy requires constant vigilance within its own system. In the context of cybersecurity, this vigilance extends to individual devices connected to the network - the endpoints. That's where Endpoint Detection and Response (EDR) enters the scene.

EDR solutions are like the intelligence officers of a network, patrolling the cyber environment, keenly observing endpoint behaviour, collecting data, and making informed decisions in real-time. They do not just build walls; they are actively looking for intruders who may have breached the walls unnoticed. From workstations to mobile devices, any entity interacting with your network is under the watchful eye of EDR.

The evolving landscape of cyber threats has outpaced traditional antivirus methods. Advanced threats can bypass these defences, covertly infiltrating systems and lying in wait to execute their malicious objectives. EDR solutions are designed to counter these advanced threats, providing an additional layer of defence by continuously monitoring endpoints for suspicious activities.

By focusing on endpoint behaviour rather than relying solely on known threat signatures, EDR solutions can uncover stealthy attacks that other defences might miss. The ability to detect anomalies in real-time allows swift response to mitigate potential damage.

But EDR is more than just a detection system. It also equips organisations with tools to respond effectively to an attack. These could range from isolating affected systems to prevent lateral movement, to providing forensic capabilities to understand the attack's origin and methodology.

Understanding this, it is clear why EDR has become a cornerstone of modern cybersecurity. It provides organisations with a deeper insight into their networks, a proactive stance against cyber threats, and a robust response mechanism to neutralise attacks swiftly. Embarking on this chapter, we will explore the nuances of EDR, the components that make it work, and the considerations when implementing an EDR strategy in an organisation.

4.6.1. Understanding EDR: Capabilities and Benefits

In the dynamic cyber environment, Endpoint Detection and Response (EDR) emerges as a robust solution that strives to turn the tables in favour of defenders. To truly appreciate the potential of EDR and its place in the cybersecurity landscape, it's necessary to delve deeper into its capabilities and the benefits it delivers to organisations.

EDR's most distinctive feature is its constant vigilance over endpoints. The system collects and records data in a centralised database from numerous points in an organisation's network. This dataset includes process executions, modifications in the registry, network connections, and more. By having access to this extensive endpoint data, EDR offers comprehensive visibility into the endpoint environment and its ongoing activities.

The crux of EDR is its analytical prowess. It employs advanced analytics, machine learning, and behavioural analysis to differentiate between normal and anomalous behaviour. This behavioural approach empowers it to detect

145

subtle indicators of a potential attack, such as lateral movement, credential dumping, or unusual data access. Unlike traditional defences that mainly rely on known threat signatures, EDR can uncover new or customised threats.

EDR solutions also come equipped with automated response capabilities. When a threat is detected, the system can respond in real-time, limiting the extent of potential damage. The responses can vary from simply sending an alert to the security team, to terminating malicious processes, or even isolating infected endpoints from the network.

While detection and response are key, another crucial capability of EDR is threat hunting. It is the proactive search for advanced threats that may be lurking undetected within the network. Security teams can use the detailed data provided by EDR to formulate hypotheses and actively look for patterns or anomalies indicative of a threat.

A significant benefit of EDR systems is the valuable insights they provide for incident response. The forensic data collected aids in understanding the attack's scope, the threat actor's path, and the overall impact of the incident. This information is pivotal not only for incident recovery but also for strengthening the organisation's defences against future attacks.

EDR solutions deliver a wealth of benefits to organisations. With their advanced detection capabilities, they provide a proactive security posture. Their ability to respond in real-time minimises the dwell time of attackers and hence, the potential damage. The in-depth visibility and analytics capabilities empower organisations to spot and rectify security gaps. Finally, their support for threat hunting activities fosters a proactive approach to security, positioning organisations one step ahead of the attackers.

On this journey of comprehending EDR, it becomes apparent that this technology is not just an added layer of security. It is a transformative approach to endpoint security, marking a significant stride in the continuous quest for robust, effective cybersecurity. The sections that follow will guide you deeper into the EDR world, explaining how it works and providing insights to help you leverage EDR effectively in your organisation.

4.6.2. Selecting an EDR Solution: Key Considerations

An organisation's choice of an Endpoint Detection and Response (EDR) solution can significantly influence its ability to safeguard its digital environment. This decision demands a thoughtful examination of multiple factors that extend beyond basic considerations of price or brand reputation. By focusing on a set of key criteria, organisations can optimise their selection process and adopt an EDR solution that aligns with their needs and aspirations.

The first significant factor to consider is the scope of endpoint coverage. With the proliferation of remote work and bring-your-own-device (BYOD) policies, employees often use a diverse range of devices and operating systems. To maintain comprehensive protection, the EDR solution should support all these various endpoints, including mobile devices and servers, across different operating systems.

A crucial aspect of an EDR solution is its detection capabilities. How efficiently can it identify both known and unknown threats? Does it use advanced techniques like machine learning and behavioural analysis to discern suspicious activities? An effective EDR solution should be able to detect a wide array of threats, from common malware to sophisticated, fileless attacks.

Real-time response capabilities are another essential attribute. Upon identifying a threat, the EDR solution must swiftly take pre-defined actions to contain it. The speed of response can drastically reduce the potential damage an attack can inflict. Additionally, the ability to customise these response actions according to the organisation's needs and risk appetite can be immensely valuable.

Visibility into endpoint activities is one of the main advantages of an EDR solution. Therefore, the solution should offer an intuitive, comprehensive dashboard that consolidates and presents data in an easily digestible manner. This capability facilitates swift decision-making during incident management and promotes better understanding of the security posture.

Interoperability with other security tools in the organisation's cybersecurity infrastructure is another crucial consideration. EDR solutions do not exist in a vacuum; they need to work seamlessly with other tools such as Security Information and Event Management (SIEM) systems, firewalls, and threat intelligence platforms. This synergy can enhance the overall security efficacy and streamline incident response processes.

The task of managing an EDR solution can be complex, especially for organisations with limited cybersecurity resources. Hence, ease of management and the availability of vendor support are important aspects to consider. Some EDR solutions also provide managed services, taking the pressure off the in-house security team and ensuring expert management of the system.

Finally, the total cost of ownership (TCO) of the EDR solution must be considered. This includes not only the purchase price but also the costs of implementation, training, and ongoing management. A thorough cost-benefit analysis can aid in choosing a solution that offers robust protection without straining the organisation's resources.

Selecting an EDR solution is a significant decision that has long-term implications for an organisation's cybersecurity health. This task requires a careful examination of the organisation's unique needs and the potential offerings in the market. The ultimate goal is to adopt a solution that strengthens the defence mechanisms, promotes proactive threat hunting, and fosters a robust security culture within the organisation.

CHAPTER 5. NETWORK SECURITY

Think of a network as a bustling city, teeming with activities and data transactions. In this bustling digital metropolis, the security measures in place form the police force, traffic signals, and even the city planners, ensuring that all activities occur in an orderly, safe manner.

However, this isn't just a simple city. It's an ever-evolving, dynamic metropolis with new residents, buildings, and roads appearing overnight, demanding that the security measures keep pace and adapt. The continuous growth of complex networks, a by-product of our increasingly connected world, necessitates robust, scalable, and adaptable network security strategies.

Network security serves as the first line of defence in protecting organisational information. It focuses on preventing and mitigating unauthorised access, misuse, malfunction, modification, destruction, or improper disclosure of networks and network-accessible resources. The importance of network security is further magnified by the surge of emerging technologies such as IoT and 5G, which significantly expand the network periphery and introduce new vulnerabilities.

This chapter delves into the intricate landscape of network security, exploring its fundamental concepts, varied strategies, key tools, and prevalent threats. This is not merely a theoretical journey, but an expedition that aims to equip you with practical insights to navigate the challenges of network security.

Network security is much like a multi-dimensional jigsaw puzzle. Each piece – be it network security policies, access controls, firewalls, intrusion detection systems, network segmentation, or encryption – holds significance in creating a complete picture of a secure network. Understanding each element's unique role, how they interconnect, and their potential impact on the overall security posture is essential to mastering the art of network security.

This chapter also sheds light on the strategies that can fortify network security. These strategies aren't about applying a one-size-fits-all solution; instead, they involve a careful assessment of the network, understanding its unique demands, and then choosing the right combination of tactics and tools. From the age-old principle of defence-in-depth to the emerging trend of zero-trust network architecture, we'll dissect the core philosophies that guide effective network security practices.

While the emphasis on proactive measures is key, we cannot turn a blind eye to potential threats. A deep understanding of possible network attacks, their mechanisms, and consequences, is an integral part of network security. This chapter will expose you to the dark underbelly of network security, exploring a range of threats from denial-of-service attacks to advanced persistent threats, and offering insights into their mitigation.

In the ever-changing world of cybersecurity, where new threats are constantly emerging and old ones evolving, staying updated is crucial. Thus, this chapter not only builds on the foundation of established knowledge but also aims to keep you abreast of the latest developments, trends, and innovations in the sphere of network security.

Embarking on this journey into the heart of network security, we invite you to cultivate a mindset that goes beyond seeing network security as a necessary evil or a compliance requirement. Instead, envision it as a critical enabler – one that safeguards operational continuity, protects data integrity, bolsters customer trust, and ultimately drives business success.

5.1. Network Protocols and Their Vulnerabilities

The realm of network security begins with a deep understanding of the fundamental building blocks of a network. These blocks are composed of network protocols that facilitate communication within and between networks. As essential as they are, network protocols also have their own set of vulnerabilities, which can become gateways for cyber threats if not addressed appropriately.

Network protocols are akin to traffic rules, governing how data moves around the network. Each protocol has a specific function, from stipulating how data should be packaged and addressed (like the Internet Protocol) to facilitating reliable data transmission (like the Transmission Control Protocol). More complex functions such as secure data transmission are also orchestrated by specific protocols such as the Secure Sockets Layer (SSL) and Transport Layer Security (TLS).

However, like any rules or instructions, network protocols are not immune to exploitation. Their vulnerabilities can be seen as gaps in the rules or oversights in their design, and these can provide opportunities for threat actors to disrupt network operations, intercept sensitive information, or gain unauthorised access.

Understanding these vulnerabilities is akin to being aware of the weak points in a fortress's defences. The walls of a fortress might be solid and high, but an unguarded postern gate or a hidden tunnel can provide an easy entry for invaders. Similarly, network protocols, despite their essential role in maintaining the functionality of the network, can become weak links if their vulnerabilities are not addressed.

The vulnerabilities of network protocols can be categorised into design flaws and implementation flaws. Design flaws are inherent in the protocols themselves, while implementation flaws arise from errors in how protocols are implemented or configured.

Protocol vulnerabilities range from those that can lead to Denial of Service (DoS) attacks, such as TCP/IP vulnerabilities, to those that can expose sensitive information, like weaknesses in wireless protocols. In some cases, older, deprecated protocols like Secure Sockets Layer (SSL) v2 and v3, which have known security flaws, may still be in use in some networks, opening them up to potential exploitation.

The challenges of dealing with protocol vulnerabilities are numerous, but not insurmountable. From patching software to following best practices in protocol configurations, multiple strategies can be employed to mitigate these risks. Moreover, awareness about these vulnerabilities among network administrators and users is a key factor in maintaining network security.

In the subsequent sections, we will delve deeper into some of the most commonly used network protocols, their known vulnerabilities, and strategies to manage these vulnerabilities effectively. This journey will not only equip you with the knowledge to identify weak points in network security but also empower you to turn these vulnerabilities into strengths. The aim is not merely to prevent attacks, but to create a resilient network environment that can adapt, recover, and learn from them.

5.1.1. Common Network Protocols and Their Functions

These protocols are the silent orchestrators of data exchange, operating under the surface to ensure seamless communication between devices in a network.

At the heart of network protocols is the Internet Protocol (IP), which serves as the fundamental layer of the internet. IP is responsible for routing packets of data from the source to the destination based on IP addresses. It's like the postal service of the internet, making sure that data gets to where it's supposed to go. However, IP itself does not guarantee that the packets will be delivered in order or even that they will be delivered at all.

That's where the Transmission Control Protocol (TCP) comes in. TCP takes care of creating a reliable connection between the sender and the receiver. It organises data packets in the correct sequence, asks for retransmission if packets are missing, and ensures that data arrives intact at the destination. This pairing of TCP and IP forms the backbone of the internet and is often referred to as TCP/IP.

On the other hand, the User Datagram Protocol (UDP) is another transport protocol that works with IP but operates differently from TCP. Unlike TCP, UDP does not establish a connection before data transmission and does not guarantee delivery or order. This makes it faster and more efficient for services that prioritise speed over reliability, like live streaming or online gaming.

In terms of securing the data during transmission, Secure Sockets Layer (SSL) and its successor, Transport Layer Security (TLS), are widely used protocols. They encrypt the data between the client and the server, ensuring that sensitive information like credit card numbers or login credentials remains confidential. These protocols are typically used in HTTPS, which is the secure version of HTTP used for web browsing.

Networks also have a range of other protocols that perform specific functions. The Hypertext Transfer Protocol (HTTP) is used for transmitting webpages over the internet. File Transfer Protocol (FTP) is used for transferring files between a client and a server. Simple Mail Transfer Protocol (SMTP) is used for sending emails, while Post Office Protocol (POP) and Internet Message Access Protocol (IMAP) are used for receiving emails.

On a local network level, protocols such as the Dynamic Host Configuration Protocol (DHCP) help assign IP addresses to devices on a network. Domain Name System (DNS) translates human-friendly domain names (like www.example.com) to IP addresses that machines can understand.

However, each of these protocols, while essential for network communication, have vulnerabilities that can be exploited if not properly managed. In the upcoming sections, we will delve deeper into these vulnerabilities, providing a clearer understanding of the potential risks associated with each protocol and outlining how best to mitigate these risks in a network environment. Understanding the functions of these protocols and their potential vulnerabilities is critical in maintaining the integrity, availability, and confidentiality of data in a network.

5.1.2. Identifying Vulnerabilities in Network Protocols

The concert of network protocols harmoniously working together to facilitate seamless data transmission is indeed an awe-inspiring spectacle. However, beneath this seeming perfection lurks a range of vulnerabilities that threat actors are constantly seeking to exploit. The journey to identify and mitigate these vulnerabilities is crucial to fortifying network security and protecting data.

IP, the very fabric that weaves the Internet together, is susceptible to IP Spoofing, where an attacker masquerades as a trusted host to conceal their identity or to gain unauthorised access to a network. This vulnerability can facilitate a range of malicious activities, from man-in-the-middle attacks to Distributed Denial of Service (DDoS) attacks.

TCP, the reliable courier of the Internet, is not without its flaws either. TCP Hijacking or Session Hijacking is a vulnerability where an attacker intercepts and takes over a TCP session between two machines. Since TCP is a stateful protocol, the hacker can manipulate the state, data, and sequence of the TCP connection, leading to unauthorised access or data theft.

UDP, with its speed and efficiency, is attractive for real-time applications. However, it's prone to UDP Flood attacks, a type of DDoS attack where the attacker overwhelms random ports on the targeted host with IP packets containing UDP datagrams, causing the host to repeatedly check for the application listening at that port and reply with an ICMP Destination Unreachable packet when none is found.

Secure communication protocols like SSL/TLS, which are often perceived as invincible fortresses, have also been found to be susceptible. The POODLE (Padding Oracle On Downgraded Legacy Encryption) vulnerability in SSL 3.0 allowed hackers to downgrade secure connections to exploit this outdated protocol. Similarly, the Heartbleed bug in TLS allowed attackers to read the memory of systems protected by vulnerable versions of OpenSSL software.

The HTTP protocol, which serves as the conduit for web content, is prone to attacks like Cross-Site Scripting (XSS) and Cross-Site Request Forgery (CSRF), where attackers inject malicious scripts or forge requests, respectively. Similarly, FTP, SMTP, POP, and IMAP, facilitating file transfers and email exchanges, can be exploited to perform unauthorised file transfers or send spam emails if not properly secured.

On the local network front, DHCP can be exploited by rogue DHCP servers distributing incorrect IP addresses or by DHCP Starvation attacks consuming all available IP addresses in a network. Similarly, DNS, the Internet's phonebook, is susceptible to DNS Cache Poisoning, where DNS cache data is corrupted to redirect requests to malicious sites.

The vulnerabilities inherent in these network protocols emphasise the critical importance of implementing robust network security measures. These range from routinely updating and patching protocol software, employing strong authentication methods, to encrypting data during transmission. By understanding these vulnerabilities, we can turn potential points of failure into strengths, reinforcing the network's defences against the continually evolving threat landscape.

5.1.3. Best Practices for Securing Network Protocols

The intricate network protocol tapestry, with its susceptibility to a range of threats, is by no means impervious. However, organisations can enforce an array of best practices to help fortify network protocols against malicious entities, enhancing the security of data transmission. By incorporating these measures, organisations can transform potential protocol vulnerabilities into robust fortresses of data transmission.

One of the fundamental measures involves regular patching and updates. Protocol software developers frequently release updates to address known vulnerabilities and enhance the overall security of the protocols. By staying abreast of these updates, organisations can ensure they are not susceptible to known vulnerabilities and are leveraging the latest security features.

Employing strong, multifactor authentication methods adds another layer of security to network protocols. This practice can mitigate the risk of unauthorised access to network resources, even if an attacker manages to

157

intercept network traffic. Protocols such as SSH and HTTPS, which employ strong authentication and encryption mechanisms, should replace insecure counterparts such as Telnet and HTTP.

Encryption is another essential practice in secure data transmission. By scrambling data into a format that can only be deciphered with the correct encryption key, organisations can protect the confidentiality and integrity of their data, even if intercepted. Protocols supporting strong encryption, like TLS for securing web traffic or WPA3 for wireless networks, should be used wherever possible.

Network segmentation and isolation can also significantly enhance protocol security. By segregating networks into separate segments, organisations can limit the propagation of a potential attack. Furthermore, by isolating sensitive data or systems on separate network segments, organisations can better protect critical resources.

Security configuration reviews and audits should be conducted regularly to identify misconfigurations or outdated security settings in protocol software. This proactive measure can detect potential security issues before they can be exploited by attackers.

Intrusion Detection Systems (IDS) and Intrusion Prevention Systems (IPS) can be used to monitor network traffic for signs of malicious activity. These tools can identify unusual patterns in protocol usage that may signify an attack, allowing for immediate response.

Implementing a robust firewall strategy is essential for controlling inbound and outbound network traffic based on predefined security rules. Firewalls act as gatekeepers, protecting internal network resources from potentially harmful external traffic.

Finally, it's crucial to provide regular security training for network administrators and users. Human error is often a significant factor in successful network attacks. By educating people about the safe use of network protocols, organisations can minimise the risk of inadvertent security breaches.

In the grand scheme of network security, safeguarding network protocols is a critical aspect. It's not merely about setting up defences but fostering an understanding and respect for the intricacies of these protocols. By employing these best practices, organisations can navigate the complex landscape of network protocols confidently, keeping their data secure while reaping the benefits of interconnected systems.

5.2. Securing Wireless Networks

As we become more connected, the ubiquity of wireless networks continues to surge. These invisible highways carry data from our fingertips to the farthest corners of the globe, eliminating the need for physical boundaries and opening new avenues of opportunity. However, with such great advantages comes an increased need for vigilance. The very characteristic that makes wireless networks appealing - their openness - also makes them vulnerable. This chapter will delve into the heart of wireless network security, investigating the potential risks and exploring the tactics employed to create a secure wireless environment.

As the physical restrictions fade away, wireless networks face a unique set of security challenges. They operate through radio waves, a shared medium accessible to any device within range. Therefore, not only do they have to deal with the usual security concerns such as data integrity and confidentiality, but they must also contend with issues like signal interference, unauthorised access, and rogue network devices.

These challenges mean that securing a wireless network isn't as straightforward as locking a door or erecting a firewall. It requires a blend of technologies and practices to ensure that the data travelling through the airwaves reaches its intended destination unscathed. While the threats may seem pervasive, with the right security measures in place, these digital skyways can be just as secure as their wired counterparts.

In this section, we'll explore how technologies such as Wireless Protected Access (WPA) and its iterations contribute to the creation of a secure wireless environment. We'll unravel the complexities of wireless encryption and the role it plays in safeguarding data. Alongside the technical aspects, we'll look at the best practices and policies that help mitigate risks, from simple steps like changing default passwords to more comprehensive strategies such as regular network audits and intrusion detection systems.

Wireless networks are an integral part of our modern world, connecting us in ways we never thought possible. Understanding the risks associated with their use and the measures necessary to mitigate these risks is an essential part of maintaining security in this ever-evolving landscape. This exploration of wireless network security will provide the tools necessary to ensure that these modern marvels of communication can be both powerful and secure.

5.2.1. Common Threats to Wireless Networks

Wireless networks, like any other technology, are not immune to threats. With the advantage of connectivity without the constraint of physical cables comes the reality of various forms of attack. It is crucial to understand these threats to be able to establish protective mechanisms that can detect, prevent, and respond to them. This section is a journey through the common threats that wireless networks encounter and how they potentially compromise data security and privacy.

A key threat facing wireless networks is unauthorised access, made possible because radio frequency (RF) signals are hard to contain. With the right tools, an intruder located within the range of a wireless network can attempt to gain access. These unauthorised users can consume network bandwidth, steal sensitive data, or launch attacks against the network and its connected devices.

Another significant threat is the deployment of rogue access points. These are wireless access points installed on a network without the network administrator's consent. They offer an easy entry point for attackers, often bypassing the network's established security measures. The clandestine nature of rogue access points makes them difficult to detect, emphasising the importance of regular network monitoring and audits.

Eavesdropping or sniffing is another common threat in wireless networks. In this type of attack, the malicious actor intercepts network traffic using a wireless sniffer. With the intercepted data, they can gain access to sensitive information such as login credentials, personal data, or confidential business information.

Man-in-the-middle (MitM) attacks are also prevalent in wireless networks. In these attacks, the intruder intercepts communication between two parties and can potentially alter the data without the victims realising their communication is compromised.

Denial of Service (DoS) attacks aim not at stealing information, but at interrupting or completely blocking the network's operation. The attacker overwhelms the network with traffic or signals, causing legitimate users to experience slow speeds or inability to connect.

Lastly, wireless networks also face the threat of RF jamming. Here, the attacker uses a device that emits a signal powerful enough to drown out the network's signal, causing disruption or total loss of connectivity.

Understanding these common threats to wireless networks forms the basis for effectively securing them. The following sections will delve into the solutions, tactics, and technologies used to counter these threats and ensure the resilience of wireless networks. By knowing what we're up against, we can construct an effective defence, ensuring the freedom of wireless connectivity without sacrificing security.

5.2.2. Securing Wi-Fi Networks: WPA3 and Beyond

As the reliance on wireless networks grows, the constant evolution of security standards that protect them becomes ever more critical. The heart of wireless network security today beats strongly with Wi-Fi Protected Access 3 (WPA3), the latest protocol aimed at making wireless connections more secure.

WPA3 was introduced by the Wi-Fi Alliance in 2018, a response to growing concerns over security vulnerabilities in its predecessor, WPA2. It includes significant improvements that make it more difficult for attackers to crack network passwords, offer better security on public networks, and simplify security for devices with limited or no display interface.

WPA3's robust password-based authentication, known as Simultaneous Authentication of Equals (SAE), effectively thwarts attempts at offline dictionary attacks, where an attacker guesses a network's password. With SAE, even if an attacker captures data from your network, they cannot make multiple password guesses offline – every guess requires interaction with the network, drastically slowing down attempts at forced entry.

Wi-Fi networks are often used for both personal and business purposes, making the privacy of open networks critical. WPA3 addresses this through individualised data encryption. This feature provides each user with their own encryption keys even when they are on an open network, such as a cafe's Wi-Fi, which significantly reduces the risk of Man-in-the-Middle attacks.

WPA3 also brings a feature called Easy Connect, making it safer and simpler to add devices with limited or no display interface, like smart home devices, to the network. By scanning a QR code with a smartphone or tablet, devices can be securely added to the network.

But cybersecurity is an arms race, and the next generation of security technologies is already on the horizon. Next-generation technologies include machine learning algorithms for anomaly detection, zero-trust network architectures that question every access request regardless of where it originates, and quantum encryption technologies, which leverage the principles of quantum mechanics to create virtually unbreakable encryption.

These new technologies, however, are not without challenges. For instance, machine learning algorithms require extensive training data to be effective, and the zero-trust architecture requires significant shifts in how network security is currently approached. Quantum encryption, while promising, is still in its early stages of development and requires considerable resources.

Implementing the WPA3 protocol and looking forward to the potential of new technologies underscores the necessity of an ongoing commitment to security. Securing wireless networks is not a one-time effort but a continuous process of adaptation and improvement. As we continue to rely more on wireless networks for daily activities, the importance of robust security measures will only grow.

5.2.3. Wireless Network Monitoring and Intrusion Detection

Given the ubiquitous nature of wireless networks and the inherent vulnerabilities that come with them, monitoring these networks for suspicious activity becomes an essential part of a comprehensive cybersecurity strategy. This is where wireless network monitoring and intrusion detection systems (IDS) come into play. These tools help identify, alert, and sometimes prevent unauthorised access and security breaches.

Wireless network monitoring is an ongoing process of observing and examining network activity. It involves the collection and analysis of network performance data and traffic patterns. This data provides insight into network health, allowing network administrators to proactively resolve issues that could lead to service disruptions or potential security vulnerabilities. Network monitoring can identify unusual traffic loads, irregular network connections, and sudden changes in network performance - all possible indicators of a security incident.

Intrusion detection systems, on the other hand, focus on identifying possible incidents, logging information about them, attempting to stop them, and reporting them to security administrators. IDS can be network-based, monitoring network traffic for signs of malicious activity, or host-based, monitoring important operating system files for unauthorised changes. These systems use signatures of known attacks or behaviours indicative of an attack to detect threats.

However, as attackers become increasingly sophisticated, they may use tactics that do not match any known signature or employ low-and-slow attacks that gradually exploit vulnerabilities over an extended period. To address this, many IDS now incorporate anomaly-based detection, which builds a baseline of normal network activity and alerts administrators to any significant deviations. This method can detect previously unknown threats but may result in a higher number of false positives.

Wireless IDS (WIDS) specifically caters to the security needs of wireless networks. It can detect rogue access points, unauthorised logins, encryption weaknesses, and attacks against the wireless network. WIDS can be standalone hardware appliances or integrated into other network hardware.

Implementing both network monitoring and intrusion detection offers multiple lines of defence. Network monitoring provides administrators with a bird's eye view of the network's health, allowing them to catch performance issues that could signal a security concern. Intrusion detection goes a step further by looking for known signs of malicious activity and anomalies in network behaviour.

But deploying these technologies is not a set-and-forget solution. Regular updates to the IDS signatures and anomaly detection algorithms are necessary to keep pace with evolving threats. Furthermore, due to the potential for false positives, especially with anomaly-based detection, organisations must have a well-planned incident response strategy in place. This includes trained personnel who can accurately interpret IDS alerts and take appropriate action.

By combining robust wireless network security measures such as WPA3 with comprehensive monitoring and intrusion detection, organisations can create a formidable defence against most wireless network threats. In a world increasingly reliant on wireless connectivity, securing these networks remains a top priority.

5.3. Virtual Private Networks (VPNs)

Organisations across all sectors have teams working remotely, and individuals require secure access to their digital resources while on the go. This calls for solutions that not only provide connectivity but also ensure the security and privacy of the data being transferred. This is where Virtual Private Networks (VPNs) come into play.

A Virtual Private Network (VPN) provides an encrypted connection or tunnel over a less secure network, typically the internet. It allows for the private transmission of data between the user and the network, effectively extending a private network across a public one. In doing so, a VPN safeguards your data, ensuring that all your online activities, from web browsing to sending emails, are secure from eavesdroppers.

VPNs have emerged as an essential tool in today's connected world, where data privacy and security face constant threats. With the rising number of cyber-attacks and data breaches, it's more crucial than ever for businesses and individuals alike to protect their sensitive information. VPNs provide this protection, shielding your data from unwanted viewers, and keeping your online activities private.

Organisations employ VPNs for a secure and reliable way to connect remote employees to their corporate network. With a VPN, employees can access the organisation's resources from anywhere, just as they would if they were physically present in the office. This functionality is especially valuable in our current era, with work-from-home and digital nomad lifestyles becoming increasingly common.

For individuals, VPNs offer a multitude of benefits. Apart from securing their data, VPNs can also provide some level of anonymity as they can mask the user's IP address. This can be particularly useful for individuals in regions with internet censorship, as VPNs can help bypass geographic restrictions and access blocked content.

VPNs are also widely used in securing data transmissions in IoT (Internet of Things) implementations, where data security is of paramount importance. As more devices become interconnected, the risk surface for potential attacks expands. VPNs can help mitigate this risk by providing a secure pathway for these devices to communicate.

Despite their benefits, VPNs are not without challenges. Not all VPN services are the same, and choosing a reputable VPN provider is essential to ensuring your data's privacy and security. Some free or low-cost VPN services may compromise your data privacy, while others might not provide the level of security needed for certain tasks.

In this section, we will dive deeper into the workings of VPNs, explore their benefits, and also touch upon the considerations one needs to keep in mind when choosing a VPN solution. In an age where data has become an invaluable resource, understanding and utilising tools like VPNs for data protection has become a necessity, rather than a luxury.

5.3.1. Understanding VPNs: Privacy and Security Aspects

Virtual Private Networks (VPNs) serve as a critical tool for privacy and security in the current digital landscape. These powerful tools enable users to create a secure connection over public networks, providing a line of defence against potential cyber threats. Before we dive into the technicalities, it's important to understand why the privacy and security aspects of VPNs are integral to both individuals and organisations.

The primary role of a VPN is to safeguard data and maintain privacy in an environment where digital information is often vulnerable. Without a secure connection, data transmitted over the internet can be intercepted by unauthorised entities, posing a significant risk. VPNs protect this data by creating an encrypted tunnel for the information to travel through, thereby shielding it from potential eavesdroppers.

Privacy is another crucial aspect of VPNs. Each device connected to the internet has a unique identifier called an IP address. This address can be used to track online activities, location, and even infer personal preferences. A VPN masks the user's real IP address and substitutes it with one from its own network, thus providing a degree of online anonymity.

Businesses, in particular, benefit greatly from VPNs. Companies frequently deal with sensitive information that, if compromised, could lead to significant losses. VPNs enable remote workers to securely access the company's internal network, ensuring that sensitive data remains secure even when accessed from insecure locations. VPNs also allow businesses to connect multiple networks securely, which is particularly useful for multinational corporations that need to share data across different locations.

Beyond privacy and security, VPNs offer several other benefits. For instance, they can overcome internet censorship and geo-restrictions. By masking the user's IP address and making it appear as if they're browsing from a different location, VPNs can allow access to content that would otherwise be blocked or restricted in their region.

However, it's essential to note that while VPNs provide a layer of security and privacy, they are not an all-encompassing solution. Users must remain vigilant about other aspects of their digital security, such as using strong, unique passwords and keeping their devices updated with the latest security patches.

Furthermore, not all VPNs are created equal. Various providers offer different features, security protocols, and privacy policies. Some may log user activity, which could potentially be accessed by third parties. Hence, it's crucial to carefully evaluate VPN providers based on their commitment to user privacy and the security features they offer.

As our digital footprints continue to grow, understanding the privacy and security aspects of tools like VPNs becomes increasingly important. VPNs offer a robust solution to many of the security issues inherent in today's digital landscape, providing a shield against many common cyber threats. Through the following sections, we will explore in detail the mechanisms that make VPNs such a pivotal tool in maintaining privacy and security online.

5.3.2. Implementing VPNs: Considerations and Best Practices

Implementing Virtual Private Networks (VPNs) is an essential step for both individuals and organisations to improve their online privacy and security. However, this process isn't as straightforward as it might seem. It involves several considerations and the application of best practices to ensure that the VPN can effectively provide the security that users need. In this section, we'll delve into the key elements to consider when implementing VPNs and highlight the best practices to adhere to.

The first consideration when implementing a VPN is to determine the specific needs of the user or organisation. This encompasses understanding the kind of data that needs protection, the number of users who will use the VPN, and the types of devices that will be connected. This information is essential as it influences the choice of VPN, its configuration, and the supporting infrastructure.

Secondly, selecting the right VPN service is crucial. With a myriad of VPN providers in the market, it's vital to assess each based on critical parameters such as encryption standards, privacy policies, server locations, speed, and compatibility with various devices. It's also advisable to opt for a VPN that has a 'no-logs' policy, meaning it doesn't record any information about your online activities.

A significant aspect to evaluate is the VPN's encryption protocol. Protocols such as OpenVPN, IKEv2, or WireGuard are commonly used and provide a good balance between security and speed. However, depending on the specific requirements, one might be preferable over the others. For instance, OpenVPN is highly versatile, working on various types of networks and offering a high degree of security. On the other hand, WireGuard is renowned for its speed and simplicity while still providing strong encryption.

Once the VPN is chosen, the next step is its proper configuration. Misconfigurations can make the VPN less effective or even entirely useless. Therefore, it's crucial to ensure that settings such as kill switches (which disconnects the internet if the VPN connection drops), DNS leak protection (which prevents requests from being sent outside the VPN tunnel), and the correct encryption protocol are enabled.

Moreover, regularly updating the VPN software is a best practice that should not be overlooked. Updates often come with patches for vulnerabilities that could be exploited by malicious actors. Therefore, keeping the VPN software updated is crucial for maintaining a secure VPN connection.

Another best practice is to train users, especially in an organisational context, on how to use the VPN correctly. This includes educating them on why they should always connect to the VPN when accessing corporate resources, the risks of connecting to public Wi-Fi without a VPN, and how to check if the VPN is connected and functioning correctly.

Finally, monitoring and maintenance are crucial aspects of VPN implementation. Regular checks should be conducted to ensure the VPN connection is stable and secure. This may involve monitoring for any unusual activities, auditing the VPN's security, and making necessary adjustments to improve its effectiveness.

Implementing VPNs requires a blend of strategic planning, careful selection, proper configuration, continuous monitoring, and consistent user education. By considering the points discussed in this section, users and organisations can harness the full potential of VPNs in protecting their online privacy and securing their data against potential cyber threats.

5.4. Network Segmentation and Isolation

several strategic and technical approaches have been devised to enhance the security posture of both enterprise and personal networks. One such potent strategy is network segmentation and isolation, which has proven to be an effective measure in bolstering network security and mitigating the potential impact of cyber threats.

Network segmentation involves dividing a network into smaller, separate segments or subnets, each acting as its individual network. Each segment can then be controlled and secured independently. This partitioning approach is analogous to the compartments of a ship. If one compartment gets breached, the water is confined to that particular section, preventing the entire ship from sinking. Similarly, if a cyber threat compromises one segment of the network, the damage can be contained, preventing it from spreading to the entire network.

On the other hand, network isolation takes the segmentation concept further by ensuring that these individual network segments cannot communicate with each other without specific authorisation. It adds an additional layer of security by reducing the 'attack surface'— the areas vulnerable to potential threats. By isolating different parts of the network, you reduce the paths that an attacker can take, making it more difficult for them to navigate through the network and reach sensitive data.

Network segmentation and isolation are not one-size-fits-all solutions; their implementation should be carefully designed based on the specific needs and structure of an organisation. It's essential to map out the network's assets, understand their interactions, and identify the critical segments that require the most protection. Commonly, organisations segregate their networks based on functions or departments, like finance, human resources, or IT. They may also segregate networks based on sensitivity levels of the data held, creating more robust barriers around high-value or high-risk data.

Once the segments are defined, security policies should be established for each of them. These rules dictate what kind of traffic is allowed in and out of a segment and should be as restrictive as possible. For example, if a segment is designed to support a database server, the policies should only allow database-related traffic and block all unnecessary services.

In addition to enhancing security, network segmentation and isolation have several other benefits. They can improve network performance by limiting unnecessary traffic and broadcast noise. They can also simplify network management and troubleshooting, as issues can be isolated and addressed in a specific segment without affecting the entire network.

However, it's important to note that while network segmentation and isolation add significant security value, they're not a panacea for all network security threats. They should be part of a broader, layered defence strategy that includes other security measures like firewalls, intrusion detection systems, and regular security audits.

In conclusion, network segmentation and isolation are powerful strategies for enhancing network security. With careful planning and implementation, they can significantly reduce the risk of a full-scale network compromise in the event of a cyber-attack. But as always, these strategies are most effective when combined with a comprehensive approach to network security, where multiple layers of defence work together to protect against a broad spectrum of threats.

5.4.1. Benefits of Network Segmentation

Network segmentation, the practice of subdividing a network into separate and securable subnets, brings forth numerous advantages to an organisation's security strategy. Beyond the fundamental principle of compartmentalising a network to limit the scope of a potential breach, segmentation yields further gains that enhance overall security, operational efficiency, and regulatory compliance.

The first and perhaps the most significant benefit of network segmentation is the increase in overall network security. By compartmentalising the network, an attack is generally constrained to the compromised segment, thereby limiting its scope and potential damage. By isolating the breach, IT teams have the opportunity to rapidly address and neutralise threats without the risk of it spreading to other parts of the network. This segmentation, therefore, acts as a critical line of defence, reducing the lateral movement of threats across the network.

Improved performance is another significant advantage of network segmentation. When a network is divided into smaller segments, there is a reduction in the volume of traffic each segment must handle, which minimises congestion and enhances the overall speed and efficiency of the network. For instance, separating bandwidth-intensive departments or operations into different segments can prevent these activities from slowing down the rest of the network.

The enhanced visibility and control over network traffic is another beneficial aspect of network segmentation. It allows for better monitoring of data flows within each segment, enabling quicker identification and response to unusual or suspicious activities. This granularity can accelerate incident response times and make the troubleshooting process more efficient and manageable.

Network segmentation also facilitates regulatory compliance, a key concern for many organisations. Several regulatory frameworks, such as the Payment Card Industry Data Security Standard (PCI DSS), require specific types of data to be isolated from other network segments. By segregating networks, organisations can ensure that sensitive data is adequately protected and meets compliance requirements.

Another indirect benefit is cost efficiency. Network segmentation allows different security policies and controls to be applied to different network segments based on their specific needs and risk profiles. This tailored approach means that resources are not wasted implementing robust security measures in low-risk areas, allowing for more effective allocation of security investments.

Furthermore, network segmentation promotes the principle of least privilege, granting users access only to the resources they require to fulfil their responsibilities. This segregation can significantly reduce the risk of insider threats, as individuals can no longer access all the resources on a network by compromising a single set of credentials.

5.4.2. Implementing Network Segmentation: Strategies and Guidelines

Implementing network segmentation can be a complex but worthwhile endeavour, requiring careful planning, comprehensive understanding of the organisation's network topology, and meticulous execution. Here, we delve into strategies and guidelines to consider while undertaking this critical task, focusing on an approach that blends security needs with organisational operations, without losing sight of compliance requirements.

The first step in implementing network segmentation is creating a clear and detailed map of your organisation's existing network topology. This process involves understanding the flow of data, identifying critical assets, pinpointing communication patterns, and outlining dependencies. The goal here is to ensure you have a complete picture of your network's structure and operations to make informed decisions about how to best segment it.

Next, prioritise the segments based on risk and business criticality. Not all segments of a network require the same level of security. Some house sensitive data or critical infrastructure that if breached, would cause significant damage to the organisation. These segments should be given the highest priority. This risk-based approach helps allocate resources more effectively, applying stringent controls where they're most needed.

When designing the network segments, adhere to the principle of least privilege. This principle dictates that a user or system should only have the necessary access to fulfil its function and nothing more. This means that segments should be as granular as possible, restricting access to resources and limiting potential damage in case of a breach.

Be sure to deploy appropriate security controls at each segmentation boundary, often referred to as demilitarised zones (DMZs). Firewalls, intrusion detection and prevention systems (IDS/IPS), and access control lists (ACLs) are typical tools used here. They help ensure that only authorised traffic can traverse between segments, further bolstering the security stance.

Another crucial aspect is continuous monitoring and regular audits of the segmented network. Implementing segmentation is not a one-time activity but a dynamic process. As the organisation grows, its network and its segmentation strategy must evolve in tandem. Regular monitoring can detect anomalies that might indicate a breach, while routine audits ensure that the segmentation strategy continues to meet the organisation's needs and compliance requirements.

Don't forget to factor in potential impacts on network performance and end-user experience. Although security is paramount, the implementation

177

of network segmentation should not unduly interfere with the business operations or create significant inefficiencies. It's important to strike a balance between securing the network and maintaining business productivity.

Lastly, it's wise to invest in automation tools to manage network segmentation. As networks grow in size and complexity, manually managing network segmentation can become unwieldy. Automation tools can help streamline the process, ensuring changes are applied consistently across the entire network, reducing the potential for human error.

As you can see, network segmentation is more than just dividing a network into subnets. It's a strategic approach to cybersecurity, enhancing the organisation's resilience to threats. By carefully considering the unique needs of the organisation and implementing segmentation accordingly, businesses can achieve robust security without sacrificing operational efficiency or user experience.

5.5. Network Monitoring and Anomaly Detection

The importance of network monitoring and anomaly detection in a comprehensive cybersecurity strategy cannot be overstated. As cyber threats become increasingly complex and sophisticated, the traditional approach of securing the perimeter is no longer enough. Organisations must invest in robust, real-time network monitoring systems that can identify and respond to unusual behaviour, and possibly catch a cyber-attack in progress.

Network monitoring involves continuously observing and analysing network traffic to ensure its functioning correctly and efficiently. It involves tracking a multitude of parameters, such as network performance, utilisation rates, and uptime availability. Yet, it also has a critical role in cybersecurity. It helps to detect any abnormal activity or deviations from established norms that could be indicators of a security breach.

On the other hand, anomaly detection, a subset of network monitoring, focuses specifically on identifying patterns in network traffic that deviate from the expected behaviour. Anomalies may be caused by a variety of factors, including technical malfunctions, human error, or malicious activities like hacking or data exfiltration. The goal is to detect these anomalies as quickly as possible to minimise their potential impact on network operations and security.

Network monitoring and anomaly detection present a considerable challenge due to the sheer volume of data involved. Networks generate vast amounts of traffic data, making it difficult to distinguish between normal and potentially harmful behaviour manually. Advanced solutions often leverage machine learning algorithms to analyse traffic patterns, detect anomalies, and even predict future network events.

Effective network monitoring and anomaly detection can offer multiple benefits for organisations. They can help identify performance issues or technical faults early, preventing potential downtime or service degradation. From a security perspective, they can detect signs of intrusion, data breaches, or insider threats, enabling swift response and mitigation.

Furthermore, these practices can also aid in regulatory compliance. Many standards and regulations require continuous monitoring of networks for any security incidents. Anomaly detection systems can provide comprehensive logging and reporting tools that can assist in meeting these requirements.

However, like any cybersecurity measure, network monitoring and anomaly detection are not without their challenges. There is the risk of false positives, where normal but unusual traffic patterns are flagged as potential threats, leading to unnecessary alerts. There are also privacy concerns, as network monitoring requires access to all traffic passing through the network, including potentially sensitive information.

In the sections to follow, we will delve deeper into various aspects of network monitoring and anomaly detection. We'll explore common tools and techniques used in this practice, best practices for implementation, and how to navigate some of the challenges. We'll also look at how these processes are being transformed by advancements in artificial intelligence and machine learning, offering even greater potential for securing network environments.

5.5.1. Importance of Network Monitoring in Cybersecurity

The digitisation of operations across sectors has resulted in a deeply interconnected environment. While this interconnection has revolutionised how businesses operate, it has also made the task of securing sensitive data and systems increasingly complex. To add to the complexity, the myriad of threats in today's cyberspace necessitates continuous vigilance. It is in this context that network monitoring assumes paramount importance for effective cybersecurity.

Network monitoring in cybersecurity is the process of consistently observing and inspecting network traffic to detect signs of potential threats or vulnerabilities. This scrutiny is crucial as early detection is often the key to minimising the damage of a cyberattack. When network monitoring is effective, it enables real-time or near-real-time detection of anomalies, helping to identify issues before they become crises.

One of the primary benefits of network monitoring in cybersecurity is enhanced visibility. In a sprawling network with multiple devices, applications, and users, keeping track of all activities is a formidable task. However, network monitoring tools can provide a detailed view of the network's status, highlighting the usage patterns, performance metrics, and any anomalies in network behaviour. This transparency allows security teams to react quickly to unusual activities and reduce the window of opportunity for potential attackers.

Moreover, network monitoring aids in detecting both external and internal threats. While firewalls and intrusion detection systems are crucial in defending against external attacks, they may not suffice when threats emerge from inside the network. By monitoring user behaviour, network monitoring can help identify potentially harmful activities, such as data misuse or unauthorised access attempts, indicating insider threats.

From a broader perspective, network monitoring contributes to an organisation's risk management efforts. By identifying weaknesses and vulnerabilities in the network, it enables a proactive approach to risk mitigation. It provides valuable insights to inform decisions about resource allocation, security policy formulation, and the overall direction of an organisation's cybersecurity strategy.

Network monitoring also plays a crucial role in ensuring compliance with regulatory standards. Several laws and regulations necessitate the tracking and reporting of specific network activities, and breaches can result in significant penalties. By generating comprehensive activity logs, network monitoring can facilitate audit processes and help demonstrate compliance.

Lastly, with the increased adoption of remote work practices and the proliferation of Internet of Things (IoT) devices, the network perimeter is continually expanding, making effective security controls more challenging to implement. Network monitoring provides the necessary oversight to keep up with these changes and ensure a robust security posture.

Despite these benefits, implementing effective network monitoring is not without challenges. It requires skilled personnel, substantial resources, and a well-planned strategy. Furthermore, it requires striking a balance between security and privacy, as monitoring traffic involves access to sensitive information. In the following sections, we'll delve into the strategies, technologies, and best practices to address these challenges and make the most of network monitoring for cybersecurity.

5.5.2. Anomaly Detection: Identifying Signs of Network Intrusion

Anomaly detection is a fundamental element of network monitoring in cybersecurity. It involves using statistical methods, machine learning techniques, and heuristic algorithms to identify patterns that deviate from the normal behaviour, potentially indicating a cyber threat. Recognising the significance of anomaly detection helps in understanding its role in identifying signs of network intrusion.

In the realm of cybersecurity, network anomalies could range from subtle changes in data traffic patterns to significant spikes in network usage. Such anomalies may indicate a wide range of potential threats, such as malware infections, Distributed Denial of Service (DDoS) attacks, data exfiltration, or even unauthorised use of resources. By promptly detecting and investigating these anomalies, organisations can thwart cyber-attacks, often before they cause significant damage.

The first step in anomaly detection is to define what constitutes 'normal' network behaviour. This baseline is established using historical network data and includes various parameters such as traffic volume, request types, protocol usage, and data packet patterns. It's important to note that the 'normal' baseline is dynamic, reflecting the ever-evolving nature of networks, user behaviour, and technology trends. Consequently, regular updates and revisions are necessary to maintain an accurate and useful baseline.

Once the baseline is established, the next step is to monitor network activities and compare them against this baseline continuously. Machine learning algorithms can be employed to automate this process, capable of processing vast volumes of network data and identifying anomalies with high precision. When an anomaly is detected, alerts can be sent to the cybersecurity team for further investigation and response.

Anomaly detection also aids in uncovering zero-day threats. These are vulnerabilities unknown to software vendors or security teams, making them particularly dangerous. Traditional, signature-based security tools struggle to detect such threats as they rely on known threat signatures. In contrast, anomaly detection can identify zero-day threats by noting abnormal network behaviour, even without prior knowledge of the threat.

183

Although anomaly detection is a powerful tool in cybersecurity, it also comes with challenges. The precision of anomaly detection is critically dependent on the quality and relevance of the baseline. An outdated or inaccurately defined baseline can result in a high number of false positives, leading to alert fatigue among security teams. Moreover, sophisticated cyber-attacks can mimic normal behaviour, evading anomaly detection systems. As a result, anomaly detection should be a part of a multi-layered security strategy, complementing other security measures.

Furthermore, implementing effective anomaly detection requires technological investment and skilled personnel. Advanced analytics tools, machine learning algorithms, and secure storage for massive volumes of network data are necessary. At the same time, cybersecurity teams need to have the expertise to interpret the results accurately and respond appropriately.

In summary, anomaly detection serves as an essential component of network security, offering the ability to proactively identify potential cyber threats. As networks continue to evolve and expand, the ability to detect and respond to anomalies in real-time will be critical in maintaining a strong cybersecurity posture. In the subsequent sections, we will explore the specific techniques, tools, and best practices for implementing successful anomaly detection in network monitoring.

CHAPTER 6. COMPUTER AND DEVICE SECURITY

As we delve into the next chapter of our exploration into cybersecurity, we turn our attention to the critical realm of computer and device security. In our interconnected world, where digital devices form an integral part of our lives and businesses, ensuring the security of these devices is a necessity, not a luxury. Cyber threats have evolved in sophistication and scale, making no device - be it a personal computer, a smartphone, or an Internet of Things (IoT) gadget - immune to potential attacks. As such, a comprehensive understanding of computer and device security is essential for anyone interested in the field of cybersecurity.

In this chapter, we will explore the many facets of securing computers and digital devices. From understanding the common threats to different types of devices, to delving into the protective measures and best practices for ensuring their security, we aim to provide a broad yet detailed overview of this crucial aspect of cybersecurity.

While the term 'computer security' might evoke images of desktop PCs and laptops, it is important to acknowledge the growing relevance of other devices such as smartphones, tablets, and an array of IoT devices. Each of these devices presents its own unique security challenges, driven by the nature of their usage, the diversity of their operating systems and applications, and the sensitivity of the data they handle.

We will start the chapter by examining the current landscape of device threats. Here, we'll explore the different types of malware, the rise of mobile-specific threats, and the unique challenges presented by IoT devices. This overview will provide a basis for understanding the diverse range of potential threats that device users may face in today's digital environment.

Next, we'll delve into the various protective measures for computers and devices. These include technical solutions such as antivirus and antimalware software, firewalls, and device encryption. We will also consider the importance of regular software updates, the role of secure configurations, and the significance of backup and recovery options.

In addition, we will pay special attention to the human factor in device security. The role of user awareness and behaviour cannot be underestimated in ensuring device security. To this end, we'll explore the best practices for secure device usage and the importance of training and education in preventing security breaches.

As we progress through the chapter, we will continue to emphasise the importance of a multi-layered approach to security. As you will come to see, the security of computers and devices relies not only on robust technical defences but also on informed user behaviour and effective security policies.

In conclusion, this chapter aims to provide you with a comprehensive understanding of computer and device security, equipping you with the knowledge and skills to protect your digital devices from potential cyber threats. As technology continues to evolve, so do cyber threats, making the knowledge imparted in this chapter all the more vital and relevant. So, let's get started and delve deep into the world of computer and device security.

6.1. Operating System Security

An operating system (OS) is the heartbeat of any computer or digital device, managing hardware resources, running applications, and acting as the user's interface to the system. Its pivotal role in a device's functioning makes it a prime target for cybercriminals. As such, understanding and implementing robust operating system security is a fundamental aspect of cybersecurity.

In this section, we will focus on operating system security, exploring its critical importance, identifying common vulnerabilities and threats, and outlining key protective measures to ensure the integrity, confidentiality, and availability of systems. We will navigate through various operating systems, such as Microsoft's Windows, Apple's macOS and iOS, Linux, and Android, to understand their unique security considerations.

The conversation about OS security must begin with understanding the potential threats. Cybercriminals exploit vulnerabilities in operating systems to deploy malware, execute unauthorised code, escalate privileges, exfiltrate data, or disrupt system functionality. Viruses, worms, Trojans, ransomware, spyware, and rootkits all pose substantial threats to OS security. As we progress, we will delve into these threats and understand how they can compromise operating systems.

Next, we will take a look at common vulnerabilities within operating systems. Many of these vulnerabilities stem from software bugs, misconfigurations, outdated components, and weak security policies. We will discuss how these vulnerabilities can be identified and the significance of regular vulnerability assessments and penetration testing.

We will then pivot to focus on measures for securing operating systems. An overarching theme you will notice is the importance of maintaining up-to-date systems. Regular updates and patches are often provided by OS developers to fix known vulnerabilities, making system updates a key defensive strategy.

We will also explore the critical role of configuration management in OS security. Proper system configurations limit potential attack vectors by removing unnecessary services, setting appropriate user permissions, and implementing security controls.

Further, we will cover the use of security software such as antivirus and antimalware programs, intrusion detection systems (IDS), and firewalls. These tools play a significant role in detecting and preventing potential security breaches.

Lastly, we will delve into the role of access control and user management in OS security. Implementing the principle of least privilege (PoLP), user authentication, and account management are integral to preventing unauthorised access and maintaining system integrity.

Operating system security is a vast and complex domain. In this section, our aim is to provide a comprehensive yet comprehensible understanding of it, elucidating the challenges faced and strategies available to protect systems. As we journey through this critical terrain, keep in mind that OS security is not a one-time activity but an ongoing commitment to protect digital assets and data continually. It requires constant vigilance, regular updates, and an evolving understanding of potential threats and vulnerabilities. So, let's embark on this exploration and unlock the potential of secure operating systems.

6.1.1. Common OS Vulnerabilities and Threats

Operating systems, while sophisticated and essential for the functioning of modern devices, are not impervious to threats and vulnerabilities. They serve as a crucial junction of hardware and software, making them a lucrative target for attackers. In this section, we delve into the common vulnerabilities and threats that plague operating systems, highlighting the urgent need for robust security measures.

Our exploration begins with an understanding of what constitutes an operating system vulnerability. Essentially, a vulnerability is a flaw or weakness in a system that can be exploited by a threat actor to infiltrate the system, disrupt its operations, or carry out unauthorised activities. In the context of operating systems, these vulnerabilities can arise from various sources, including design flaws, programming errors, configuration mistakes, or outdated components.

Some of the most common vulnerabilities include buffer overflow errors, which occur when a program writes more data to a buffer than it can handle, thereby causing it to overwrite adjacent memory areas. Similarly, uncontrolled format strings can lead to exploits that allow an attacker to execute arbitrary code. Race conditions, where the timing of actions impacts how they are executed, can also lead to security breaches if not properly managed.

Apart from vulnerabilities, operating systems also face a myriad of threats. These include malware such as viruses, worms, trojans, and ransomware, which can infect a system, steal sensitive information, and disrupt operations. Rootkits represent another serious threat, enabling an attacker to gain privileged access and hide their activities from the user and other system processes.

Other threats to OS security include zero-day exploits, where an attacker takes advantage of a previously unknown vulnerability before a patch becomes available, and privilege escalation attacks, which enable a user to gain higher-level permissions on the system than initially granted. Notably, denial-of-service (DoS) and distributed denial-of-service (DDoS) attacks can overwhelm a system, rendering it inaccessible to legitimate users.

Furthermore, social engineering attacks such as phishing or baiting often target users rather than the operating system directly. By tricking users into disclosing sensitive information or downloading malicious software, attackers can circumvent otherwise robust security systems.

As the world continues to become more interconnected, the number and complexity of these threats and vulnerabilities are on the rise. It is therefore critical to comprehend the nature of these potential risks to develop effective strategies for managing them.

In this section, we have merely scratched the surface of the vast threat landscape that exists for operating systems. However, by understanding the common threats and vulnerabilities, you will be better equipped to protect your systems. In the upcoming sections, we will delve into the strategies and best practices for securing operating systems, underscoring the critical need for continuous vigilance, regular updates, and robust security protocols in the fight against cyber threats.

6.1.2. Secure Configuration of Windows, Linux, and MacOS

The configuration of an operating system plays a pivotal role in maintaining the security and integrity of a system. In a world where cyber threats persistently evolve, learning to securely configure Windows, Linux, and MacOS operating systems is crucial. This section provides insight into the process of optimising these systems to maximise security, minimise vulnerabilities, and ensure robust, efficient operations.

Starting with Windows, it is the most widely used operating system globally, making it a common target for attackers. Secure configuration begins with the principle of least privilege, where users and processes are given the minimum privileges, they require to carry out their tasks. Regular updates are another important aspect. Microsoft frequently releases patches to fix known vulnerabilities, so ensuring automatic updates are enabled is a key security measure. Windows also provides a host of built-in security features such as Windows Defender, Windows Firewall, and BitLocker, which should be properly configured and utilised to their fullest extent.

Linux, known for its flexibility and powerful command-line interface, also requires careful configuration for optimal security. Like Windows, it is important to implement the principle of least privilege. Moreover, Linux systems should be regularly updated, and unused services and software should be removed to minimise the attack surface. Tools such as AppArmor or SELinux can help enforce access controls and restrict processes. Additionally, system logging and monitoring should be enabled to detect and respond to potential threats in a timely manner.

MacOS, despite its reputation for robust security, is not invulnerable to threats. Therefore, it too requires secure configuration. Similar principles apply here – regular updates, least privilege, and utilising built-in security features. Features unique to MacOS, such as Gatekeeper, which prevents unapproved software from running, and XProtect, a form of malware detection, should be properly configured and updated. FileVault 2 can provide full disk encryption, further protecting data in case of physical theft

Beyond these specific measures, there are also common practices applicable to all operating systems. For instance, user accounts should be secured with strong, unique passwords, and two-factor authentication should be enabled where possible. Regular backups should also be part of your security strategy to protect against data loss. Network security should be ensured with firewalls and secure Wi-Fi configurations.

Keep in mind, however, that secure configuration is not a one-time activity. As new vulnerabilities are discovered and as your system and needs evolve, your configurations may need to be updated or adjusted. Regular auditing and monitoring can help identify areas of concern and keep your system secure.

In conclusion, the security of an operating system is significantly influenced by its configuration. As such, understanding how to securely configure Windows, Linux, and MacOS is key in protecting against cyber threats. This understanding, coupled with continuous vigilance, can significantly mitigate the risk of compromise, ensuring the confidentiality, integrity, and availability of your systems and data.

6.1.3. OS Security Features and How to Use Them

Operating systems are designed with a range of security features to protect the system and its data. However, these features are only effective if they are correctly implemented and used. From user account controls to firewall settings, and from system updates to encryption tools, understanding the array of security features in an operating system is the first step towards a stronger defence line against cyber threats.

Let's begin with user account controls, a common feature across Windows, Linux, and MacOS. These controls allow administrators to set the level of access for each user, implementing the principle of least privilege. They prevent unauthorised access to critical system components and reduce the potential damage if a user account is compromised.

Another vital security feature is the firewall. Firewalls act as the first line of defence against incoming threats, blocking unauthorised access while allowing outbound communication. For Windows users, the Windows Defender Firewall offers customisable settings. Linux distributions often come with iptables or nftables, while MacOS has an integrated firewall in its Security & Privacy settings.

Antivirus software is also a fundamental security feature. Many operating systems come with built-in antivirus solutions like Windows Defender Antivirus for Windows and XProtect for MacOS. Linux distributions do not usually include an antivirus by default, but plenty of reputable options can be installed.

Operating systems often provide tools for data encryption. Data encryption ensures that even if a device is physically stolen or compromised, the data it contains remains inaccessible without the encryption key. Windows offers BitLocker, MacOS provides FileVault, and Linux users can utilise the dm-crypt package or third-party tools like VeraCrypt.

Lastly, an often-underestimated security feature is the system update mechanism. Regular updates patch known security vulnerabilities and improved system stability. Windows has Windows Update, MacOS uses Software Update, and various package managers like apt, yum, or pacman handle this in Linux.

Using these security features involves understanding their functionalities, knowing how to configure them, and keeping them updated. It is essential to stay informed about the latest updates and improvements to these features, as cybersecurity is a fast-paced field where threats continually evolve. System administrators and users alike must ensure they are leveraging these tools to provide a secure operating environment.

In conclusion, operating systems offer a rich set of security features designed to protect against cyber threats. To benefit from these, users must understand their functions and configuration. Remember, these tools form the basis of your cybersecurity defences. Regular updates, combined with best practices like least privilege and data encryption, can create a robust security posture that guards your system and its valuable data against the wide array of threats in the cyber landscape.

6.2. Mobile Device Security

As we increasingly depend on smartphones and tablets to handle both personal and professional tasks, the realm of mobile device security has grown in significance. The breadth of information stored on our devices, from financial data to personal photos, and the range of capabilities they offer, from banking transactions to remote work, make them enticing targets for cybercriminals. Given this reality, understanding and implementing robust security measures for mobile devices are essential elements of a comprehensive cybersecurity strategy.

Mobile device security involves protecting both the physical device and the information it contains from various threats, which could be unauthorised access, data leakage, malware, phishing, or network-based attacks. While the nature of these threats is similar to those targeting computers, mobile devices are subject to unique vulnerabilities due to their portability, modes of connectivity (including cellular networks and public Wi-Fi), and the variety of apps they accommodate.

One defining characteristic of mobile device security is the sheer variety of operating systems, with Android and iOS being the dominant ones, but also including others like Windows Mobile and Blackberry. Each of these operating systems has distinct security features, strengths, and weaknesses. Therefore, understanding the specifics of each is crucial to enhance mobile device security effectively.

Moreover, mobile devices often have access to sensitive corporate data via work emails, cloud storage apps, and other productivity tools. This intersection between personal and professional data necessitates a careful balance between security and privacy, especially in a Bring Your Own Device (BYOD) policy context. Companies must protect their information without infringing on the personal privacy of employees, and individuals must be wary of the data they access on personal devices and the apps they install.

In the forthcoming sections, we'll delve deeper into the various facets of mobile device security. We'll look at prevalent mobile security threats and how to safeguard against them. We'll also explore the security features available in popular mobile operating systems and how to optimise them for maximum protection. For organisations, we'll delve into mobile device management strategies and best practices for implementing a BYOD policy. Mobile device security is a critical aspect of cybersecurity in today's connected world, and understanding it is paramount for both individual users and businesses.

6.2.1. Mobile Threat Landscape: Android and iOS

With the convenience and wide-ranging capabilities they offer, smartphones have become indispensable in our everyday lives. As of 2023, Android and iOS, Google and Apple's mobile operating systems, respectively, account for most of the mobile OS market share worldwide. However, the ubiquity and essential functions of these systems have made them appealing targets for cybercriminals.

The mobile threat landscape, just like its traditional counterpart, is highly dynamic and continually evolving. The spectrum of threats faced by Android and iOS users varies from those rooted in physical access to the device to complex cyber-attacks aimed at exploiting software vulnerabilities and human fallibility. Therefore, understanding the prevailing threats specific to these operating systems and the underlying mechanisms that make these attacks possible is a prerequisite for ensuring mobile security.

Android, being an open-source operating system with a more lenient app approval process, is inherently more susceptible to malware and phishing attacks. The diversity of Android device manufacturers also leads to an inconsistent rollout of security updates, leaving many users with out-of-date, vulnerable software.

Conversely, iOS is a closed-source operating system, meaning Apple strictly controls the system's code and the applications allowed in the App Store. These restrictions provide a more secure environment by minimising the risk of malicious apps. However, this doesn't entirely eliminate the threat of cyber-attacks. For instance, iOS devices are still vulnerable to network-based attacks, phishing, and exploitation of vulnerabilities present in the OS and third-party apps.

A significant portion of mobile threats capitalises on user behaviour. These attacks often involve social engineering techniques that trick users into revealing sensitive information, clicking on malicious links, or downloading malicious apps. Examples include phishing attacks via SMS (smishing) or voice call (vishing), fraudulent apps masquerading as legitimate ones, and Man-in-The-Middle (MitM) attacks on insecure public Wi-Fi networks.

The focus on the Android and iOS threat landscape is crucial because these two platforms represent the vast majority of mobile users. As we delve deeper into the nuances of mobile security threats, we'll explore how these platforms, by design or policy, deal with these threats differently. We'll look at the inherent security features and controls within these systems, how they are exploited, and what you can do as a user or an organisation to fortify your defences.

Understanding the terrain of mobile vulnerabilities is the first step in protecting sensitive data. The next sections will provide the map to navigate this terrain, highlighting the precautions that users can take to fortify their mobile devices against the prevalent threats. By understanding the threat landscape and employing smart cybersecurity strategies, we can make the most of our mobile devices without compromising security.

6.2.2. Implementing Mobile Device Management (MDM)

The prolific use of mobile devices in the workplace has introduced new security challenges. Organisations now grapple with the task of managing a myriad of devices—smartphones, tablets, laptops—that provide multiple entry points for potential security breaches. Mobile Device Management (MDM) has emerged as a key solution to manage, monitor, and secure these devices.

MDM involves the use of software to monitor, manage, and secure employees' mobile devices that are deployed across multiple mobile service providers and across numerous mobile operating systems being used in the organisation. It has become an essential tool for IT departments to manage the complex and diverse array of devices and operating systems, all while ensuring the organisation's network security.

With MDM, organisations can enforce device security policies, deploy software and apps, secure access to corporate data, track devices, and even remotely wipe data off the devices if they're lost or stolen. Implementing MDM starts with selecting a solution that suits the organisation's needs in terms of compatibility with existing systems, cost-effectiveness, ease of use, and the level of control it offers over mobile devices.

A critical step in the implementation process is the creation of a comprehensive mobile device policy. This policy should clarify what types of devices are permitted, what level of access these devices have to the corporate network, which user behaviours are acceptable, and what security measures are mandatory (such as password protection, encryption, and so on).

Once the policy is in place, the next step is to configure the MDM software to enforce these policies. This might involve setting up automated compliance checks and actions to be taken if a device is found to be non-compliant. Depending on the MDM solution chosen, it might also be possible to manage software updates and distribute apps and documents directly to the devices.

Training and awareness are also crucial elements of the implementation process. Employees need to be made aware of the policies and educated on safe mobile device usage. This includes training on avoiding phishing attempts, using secure Wi-Fi connections, and recognising signs of malware or compromised security.

Lastly, an effective MDM implementation requires regular review and updates. The rapidly evolving mobile threat landscape means that what worked a year or even six months ago might not be sufficient today. Regular audits and updates will ensure that the organisation's mobile device management remains robust and effective against emerging threats.

Implementing an MDM solution is a significant step towards ensuring mobile device security within an organisation. It provides a systematic way to manage the security challenges posed by the increasing reliance on mobile devices in the workplace. The subsequent sections will further explore the key features to look for in an MDM solution and discuss some best practices for mobile device security. The goal is to create an environment where productivity and security can go hand in hand, leveraging the advantages of mobile technology without compromising organisational data and systems.

6.2.3. Securing Mobile Applications: Best Practices

In our contemporary digital ecosystem, mobile applications serve as the backbone for a large majority of personal and professional tasks. This includes everything from checking emails and accessing work files to financial transactions and social networking. In 2021, consumers downloaded an astounding 218 billion apps, marking a consistent growth trend. As the use of mobile applications expands, so too does the potential attack surface for cyber adversaries. It is, therefore, crucial to integrate robust security practices to protect mobile applications from a broad range of cyber threats.

Securing mobile applications involves a multi-faceted approach that spans the entire lifecycle of the application, from the initial design and development phase through to post-deployment maintenance. Incorporating security at every stage ensures the application is built on a solid foundation and can withstand evolving threats.

At the development stage, one of the best practices is adhering to the principle of secure coding. This involves implementing strategies to prevent common security flaws such as SQL injection, cross-site scripting, and buffer overflows. Techniques like input validation, parameterised queries, and secure error handling can go a long way in creating a secure codebase.

Another essential practice is to encrypt all sensitive data that the app handles. Whether the data is at rest, stored in the device's memory, or in transit between the device and a server, encryption ensures that even if data is intercepted, it remains unreadable and secure. Using the latest encryption standards is vital to keeping up with advancements in decryption techniques.

A comprehensive security testing strategy is also pivotal in securing mobile applications. This should include a combination of penetration testing, vulnerability scanning, and security audits. Automated testing tools can help identify common vulnerabilities, while manual testing can offer a deeper, more nuanced analysis.

Post-deployment, mobile applications should be kept updated with patches for known vulnerabilities. This is particularly important considering that many cyberattacks exploit known but unpatched vulnerabilities. Ensuring automatic updates and timely patch management processes can drastically reduce the potential attack surface.

Furthermore, mobile applications should request only the minimum necessary privileges from the user. Over-privileged applications present a high risk because they can access more data and features of the device than necessary, which can be a goldmine for hackers. Developers should follow the principle of least privilege, meaning an app should have only the permissions necessary to perform its function and no more.

Finally, it's crucial to educate users about safe app practices. Users should be encouraged to download apps only from trusted sources, check permissions requested by apps, and regularly update apps to the latest version.

In conclusion, securing mobile applications is a complex task that demands a comprehensive, lifecycle-oriented approach. However, by adhering to best practices in secure coding, data encryption, security testing, privilege management, and user education, developers and organisations can significantly mitigate the risks associated with mobile applications.

6.3. IoT Security

The Internet of Things (IoT) is an interconnected web of physical devices embedded with sensors, software, and network connectivity, enabling them to exchange and analyse data. The IoT landscape has been expanding at a breath-taking pace, driven by advancements in sensor technology, AI, and wireless networks. From smart homes and healthcare devices to industrial sensors and autonomous vehicles, IoT technology is transforming numerous industries and our daily lives.

However, as the IoT ecosystem continues to grow, it brings along a unique set of security challenges that need to be addressed. The ubiquitous nature of these devices, coupled with their often-inadequate security measures, makes them an appealing target for cybercriminals. Additionally, the high volume of data these devices collect, transmit, and process further underscores the critical need for robust security measures.

The vast heterogeneity of IoT devices exacerbates these challenges. With a wide array of manufacturers, operating systems, and software stacks, there is no one-size-fits-all approach to IoT security. Each device class has its unique vulnerabilities and potential mitigation strategies. The multitude of IoT communication protocols – such as Zigbee, Bluetooth, Wi-Fi, and cellular – also present their unique security considerations.

Another layer of complexity is introduced by the sheer scale of IoT networks. As businesses and consumers continue to adopt more connected devices, ensuring the security of these sprawling networks becomes exponentially more challenging. This large-scale deployment of devices also presents a significant risk of botnet attacks, where numerous devices are infected and harnessed to conduct malicious activities, such as Distributed Denial of Service (DDoS) attacks.

The physical accessibility of many IoT devices presents another distinct security challenge. In contrast to servers housed in secure data centres, IoT devices are often deployed in unsecured environments, making them vulnerable to physical tampering.

Securing the IoT ecosystem, therefore, requires a holistic approach that takes into account the unique characteristics and challenges presented by these devices. This includes secure design principles, robust authentication and encryption mechanisms, regular firmware updates, secure communication protocols, and strategies to manage the risk posed by physical accessibility.

In the following sections, we'll dive deeper into the specifics of IoT security, including common threats, recommended security measures, and best practices for securing IoT devices and networks. Despite the challenges, with a proactive and comprehensive approach to IoT security, it's possible to harness the benefits of IoT technology while minimising its risks.

6.3.1. Challenges in Securing IoT Devices

As the Internet of Things (IoT) expands, the challenges in ensuring its security also escalate. The unique characteristics of IoT devices, such as their diversity, ubiquity, and connectivity, create an intricate web of complexities that make securing these devices a daunting task.

Firstly, the enormous heterogeneity of IoT devices is a substantial challenge. IoT devices range from tiny sensors to complex machinery, each with its unique hardware and software specifications. With varying operating systems, processors, memory capacities, and capabilities, developing a standardised security framework that caters to all these devices is an enormous task. This variety of devices leads to a broad attack surface with multiple potential vulnerabilities that could be exploited by cybercriminals.

Another significant challenge is the sheer number of IoT devices. With billions of devices connected worldwide and projected growth suggesting even higher numbers, managing the security of this vast network is a Herculean task. The large scale not only complicates security management but also increases the potential damage that a compromised device could inflict on the network.

The long lifespan of many IoT devices also presents a challenge. Many of these devices are designed to last for years or even decades. However, the software and security measures in place at the time of manufacture may become outdated and insufficient over time, leaving the device vulnerable. Keeping these devices updated with the latest security patches is critical but can be logistically difficult due to the sheer number of devices and their distributed nature.

Further complicating matters is the fact that many IoT devices are physically accessible and often deployed in unsecured environments. This makes them susceptible to physical attacks that could alter the device's function, extract sensitive data, or introduce malicious software.

Connectivity is another crucial aspect. IoT devices usually communicate wirelessly over various protocols, each having its own set of vulnerabilities. Ensuring secure communication while maintaining interoperability across devices is a formidable challenge.

Finally, the significant amount of data collected, processed, and transmitted by IoT devices raises privacy concerns. Safeguarding this data from unauthorised access, ensuring its integrity, and managing its privacy are critical considerations.

In summary, the task of securing IoT devices is a complex, multidimensional challenge, shaped by the diverse, ubiquitous, and connected nature of these devices. It requires a comprehensive, multi-layered approach that includes hardware and software security, secure communication protocols, data privacy measures, and physical security. A deeper understanding of these challenges can help shape robust and effective strategies to secure IoT devices against the growing cyber threats.

6.3.2. Strategies for IoT Security: From Device to Network

Securing the IoT ecosystem is a colossal task that necessitates a strategic and comprehensive approach. By implementing strong security measures at every level—from individual devices to the broader network—organisations can better protect their IoT systems from potential threats.

Starting at the device level, it's crucial to ensure that IoT devices have strong built-in security features. Manufacturers should adhere to stringent security standards during the device's design and production stages, incorporating elements such as secure boot mechanisms, data encryption, and hardware-based security features. Furthermore, devices should have the capability to receive regular firmware and software updates to patch any discovered vulnerabilities and stay aligned with emerging security trends.

User authentication is another critical aspect at the device level. Implementing strong password policies and multi-factor authentication can prevent unauthorised access to IoT devices. For IoT devices that collect sensitive data, encryption should be used both for data at rest and data in transit.

At the network level, securing communication channels between IoT devices and the network is vital. Using secure network protocols and encrypting network traffic can shield these communications from interception or alteration. Network segmentation can also be beneficial, segregating IoT devices into separate network segments to contain the potential spread of an attack.

Regular monitoring of IoT networks is another effective strategy. By deploying intrusion detection systems and security information and event management (SIEM) solutions, organisations can identify and respond to unusual network activities or patterns that may signal a security breach.

Additionally, it's essential to secure the data generated by IoT devices. Employing robust data security practices, including encryption and secure data storage solutions, can help protect this data from unauthorised access or manipulation. Privacy-enhancing techniques such as anonymisation and pseudonymisation can also help safeguard sensitive data.

IoT device management should not be overlooked. A centralised management platform can facilitate software updates, manage device configurations, and monitor device health. Regular vulnerability assessments and security audits can help identify potential weak points and verify that all security measures are functioning as expected.

Finally, all these strategies need to be underpinned by an effective security policy that clearly outlines the responsibilities, guidelines, and procedures for IoT security within an organisation. Regular training should be provided to all stakeholders to raise awareness of the potential risks and the best practices for IoT security.

In essence, securing the IoT landscape requires a combination of robust device security, secure network communications, effective data protection measures, comprehensive device management, and a strong security policy.

By adopting these strategies, organisations can greatly enhance the resilience of their IoT networks against cyber threats.

6.3.3. Future of IoT Security: Trends and Predictions

As the Internet of Things (IoT) continues to evolve and proliferate across various sectors, the domain of IoT security is poised for a period of significant advancement and change. The scope, complexity, and scale of IoT ecosystems present unique challenges that necessitate innovative solutions and forward-thinking approaches. Here are some trends and predictions that are expected to shape the future of IoT security.

1. **AI and Machine Learning in IoT Security:** With the massive volume of data generated by IoT devices, AI and machine learning are expected to play a crucial role in improving IoT security. These technologies can help automate threat detection and response, monitor devices and networks for unusual activities, and predict potential vulnerabilities or threats.

2. **Edge Computing:** The shift towards edge computing is likely to have a considerable impact on IoT security. By processing data closer to the source, edge computing reduces the need to transmit data over long distances, thereby lowering the risk of data interception or manipulation. Moreover, it allows for quicker response to potential threats.

3. **Blockchain for IoT Security:** Blockchain technology could offer a new paradigm for securing IoT devices and networks. By leveraging a decentralised and tamper-proof ledger, blockchain can facilitate secure device authentication, data integrity, and distributed trust in IoT networks.

4. **Standardisation and Regulation:** There is an increasing recognition of the need for standardisation and regulation in IoT security. Regulatory bodies and industry consortiums are expected to introduce more stringent standards and guidelines to ensure security is embedded in the design and operation of IoT devices and systems.

5. **Secure Hardware and Firmware:** More emphasis will be placed on building hardware-level security into IoT devices. This could include secure boot processes, hardware authentication, and physical tamper resistance. Additionally, ensuring firmware can be updated in a secure and timely manner will continue to be crucial.

6. **Privacy-Enhancing Technologies:** With IoT devices often collecting sensitive data, privacy will be a growing concern. Techniques like anonymisation, differential privacy, and homomorphic encryption, which allows computation on encrypted data, are likely to gain prominence.

7. **Security as a Service (SaaS):** With the complexity of IoT security, many organisations may choose to outsource security to third-party service providers. Security as a Service (SaaS) for IoT could become a significant market, providing businesses with expert support to secure their IoT ecosystems.

8. **Cybersecurity Insurance:** As the risks associated with IoT grow, the cybersecurity insurance market is expected to expand. These policies could provide financial protection against IoT-related cyber-attacks, offering a safety net for businesses.

These trends reflect the continuous evolution of IoT and the cybersecurity landscape. However, they also underline the paramount importance of proactive and forward-thinking approaches to IoT security. As IoT devices become more ingrained in our daily lives, securing them will become not just a technical challenge, but a societal imperative.

6.4. Secure Configuration and Patch Management

As we delve deeper into the domain of computer and device security in this book, it becomes increasingly apparent that a critical aspect of this field is the establishment and maintenance of secure configurations, along with an effective patch management strategy. These are fundamental to ensuring the confidentiality, integrity, and availability of systems and data, which are the cornerstone principles of cybersecurity.

Both secure configuration and patch management carry a substantial weight in shaping the overall security posture of an organisation. They represent a proactive approach towards mitigating vulnerabilities and reducing the attack surface, thereby fortifying defences against potential cyber threats.

Secure Configuration: At the heart of this concept is the understanding that every piece of technology – be it an operating system, a software application, a server, a network device, or an IoT device – comes with a multitude of settings and features, not all of which are required for a specific use case or environment. A number of these settings and features might open doors to potential security threats. Hence, secure configuration implies tailoring the system's settings in a way that only the necessary features are enabled, unnecessary ones are disabled, and every setting is optimally adjusted to achieve a balance between functionality and security.

Patch Management: This refers to the process of keeping software within an environment up to date, including operating systems, applications, and firmware on devices. Patches are pieces of software designed to update a computer program or its supporting data, including fixing security vulnerabilities and other bugs, and improving the usability or performance. Neglecting patch management can leave systems vulnerable to exploits, leading to data breaches, system downtime, and a host of other problems.

Despite the importance of secure configuration and patch management, many organisations struggle with these areas. The reasons range from lack of awareness about the latest security best practices and lack of skilled staff, to the challenges associated with managing a diverse and complex IT environment. This chapter aims to provide a detailed understanding of these crucial cybersecurity concepts.

We will be exploring the steps involved in implementing a secure configuration, including the significance of adhering to configuration baselines and hardening guidelines, and the role of automation in enforcing secure configurations. Additionally, we'll dive deep into the world of patch management, shedding light on the process of vulnerability monitoring, patch testing and validation, deployment, and verification.

Secure configuration and patch management are not one-time activities. They require ongoing attention and continuous improvement to adapt to evolving technologies, emerging threats, and changing business requirements. However, with a comprehensive understanding of these topics and a commitment to implementing best practices, organisations can significantly enhance their cybersecurity defences, thus ensuring a safer environment for their valuable data and systems.

6.4.1. Importance of Secure Configurations

Secure configurations are pivotal for maintaining an organisation's cybersecurity posture. These configurations can be viewed as a foundational measure for safeguarding against a multitude of threats and reducing the likelihood of successful attacks. They refer to the security measures implemented to protect information by managing the settings on computers, servers, applications, and network devices.

Although it might be tempting to think of secure configurations as a mere list of checkboxes that need to be ticked, in reality, they constitute a strategic component of an organisation's cybersecurity framework. They can potentially fortify an organisation's defences against various attack vectors, including malware, hacking, data exfiltration, and insider threats.

To begin with, secure configurations significantly decrease an organisation's attack surface. By limiting unnecessary functions, services, and privileges, these configurations reduce the number of potential entry points for malicious actors, thus making it harder for them to exploit the system. Moreover, by adhering to the principle of least privilege, secure configurations ensure that users and processes only have the minimum rights necessary to perform their tasks, which can further help prevent attacks and limit their impact.

Secure configurations can also enhance visibility and control over the IT environment. This can be particularly beneficial for large organisations that have complex infrastructures with numerous components. Having a clear understanding of the configurations across the infrastructure can facilitate anomaly detection, making it easier to identify potential security incidents.

Furthermore, secure configurations play a crucial role in regulatory compliance. Various regulations and standards, such as the General Data Protection Regulation (GDPR), the Payment Card Industry Data Security Standard (PCI DSS), and the Health Insurance Portability and Accountability Act (HIPAA), require organisations to implement certain configurations to protect sensitive data. Non-compliance can result in hefty fines, reputational damage, and loss of customer trust.

Importantly, secure configurations can be cost-effective. While implementing these configurations might require some initial investment in terms of resources and time, they can save organisations a significant amount of money in the long run by preventing security breaches. According to IBM's 2020 Cost of a Data Breach Report, the average cost of a data breach is $3.86 million, a substantial expense that most organisations would like to avoid.

Finally, secure configurations contribute to creating a culture of security within the organisation. They represent a tangible manifestation of the organisation's commitment to cybersecurity, which can encourage employees to adopt secure behaviours and enhance their understanding of security principles.

In conclusion, the importance of secure configurations cannot be overstated. They are an essential part of an organisation's defence-in-depth strategy, complementing other security measures like firewalls, intrusion detection systems, and antivirus software. The following sections will delve further into how to implement secure configurations and the best practices to consider.

6.4.2. The Patch Management Process: Best Practices

Patch management is an essential aspect of cybersecurity. It involves acquiring, testing, and installing multiple patches (code changes) on an existing system and managing the technologies that deal with these processes. These patches often contain security upgrades meant to rectify vulnerabilities that malicious actors might exploit. When effectively carried out, patch management helps reduce the exposure of IT resources to security risks and maintain system integrity and confidentiality. This comprehensive write-up explores the best practices involved in the patch management process.

Firstly, it is critical to maintain an inventory of all the organisation's hardware and software. A detailed inventory gives visibility into the organisation's systems, including all endpoints, software, operating systems, and firmware. This visibility is crucial as it aids in identifying which systems might require patching. Thus, the first best practice is having an up-to-date asset inventory.

Secondly, consistent monitoring for available patches is critical. Vendors often release patches when they discover vulnerabilities in their systems. Organisations should be alert to these updates. They can subscribe to vendor notifications or use automated patch management tools that keep track of patches as they become available.

Next, organisations need to prioritise patches. Not all patches need to be installed immediately. Some might need to be expedited due to the high risk they pose, while others can be deferred. Factors such as the severity of the vulnerability, the criticality of the system, and the potential impact of the patch should influence the prioritisation.

Before installing a patch, it is advisable to test it in a controlled environment. This practice helps to identify any potential conflicts or issues that could disrupt operations. Depending on the patch's complexity, testing can range from simple checks to full regression testing.

Once a patch has been tested, it can then be deployed. Organisations need to define a process for the deployment that minimises disruption to business operations. Some patches may require systems to be rebooted, which may require scheduling patch deployment outside of regular business hours.

Following deployment, verification is necessary. It ensures that the patches have been correctly installed and are functioning as expected. Any deviations should be noted and addressed immediately.

Patch management should also entail documenting every step in the process. Documentation provides a record of what was done, why, and when. This can be valuable for auditing purposes and can help troubleshoot if problems occur later.

Lastly, continuous improvement is an integral part of patch management. Regular reviews and feedback sessions can help identify areas for improvement, making the process more effective and efficient.

In conclusion, an efficient patch management process is essential in today's dynamic cybersecurity environment. By adhering to best practices, organisations can significantly reduce their exposure to cyber threats, ensure compliance with various regulations, and maintain the integrity and confidentiality of their systems.

6.4.3. Dealing with Zero-Day Vulnerabilities

The arena of cybersecurity is a battlefield that evolves every day. In this constantly changing landscape, one of the most formidable challenges that organisations face is dealing with zero-day vulnerabilities. These are vulnerabilities in software that are unknown to the parties responsible for patching or fixing the issue. The term "zero-day" refers to the fact that developers have "zero days" to fix the problem that has just been exposed, potentially leaving systems vulnerable to exploits. This discussion aims to shed light on dealing with these elusive threats.

An understanding of the nature of zero-day vulnerabilities and their potential impacts is the first step in managing them. They often remain hidden in the code, undetected by developers and users until they are exploited by cybercriminals or discovered by security researchers. These vulnerabilities can exist in any software — from operating systems to applications, and even firmware on devices. Their exploitation can lead to significant damages, such as data breaches, disruption of services, and substantial financial costs.

The challenge with zero-day vulnerabilities lies in their unpredictability and the swift action required upon discovery. Traditional security measures, like antivirus software and firewalls, might not be effective against these threats, necessitating a more proactive and layered security approach.

Firstly, it is critical to implement security practices that reduce the attack surface — that is, the number of points where an attacker can enter or extract data. This can be achieved by employing the principle of least privilege, ensuring systems are updated with known patches, deactivating unnecessary services and software, and following secure coding practices.

Secondly, deploying advanced threat protection solutions that use behavioural analysis could be beneficial. These solutions can detect unusual behaviour, such as an application trying to modify certain system files, indicating a potential zero-day attack.

Next, it is crucial to have a robust incident response plan ready. In the event of a zero-day attack, having a well-defined process can reduce reaction time and minimise damage. The plan should include identifying the compromised systems, containing the damage, eradicating the threat, recovering the systems, and learning from the incident to prevent future occurrences.

Moreover, threat intelligence plays a significant role in dealing with zero-day vulnerabilities. By staying informed about the latest threats and vulnerabilities discovered worldwide, organisations can anticipate and prepare for potential zero-day attacks.

Lastly, consider using intrusion prevention systems and exploit prevention technologies. They can help to detect and prevent attempted exploits of vulnerabilities, including those that are zero-day.

Dealing with zero-day vulnerabilities requires a comprehensive, layered approach to security. It requires ongoing vigilance, the application of robust security controls, and a proactive mindset towards potential threats. By embracing such an approach, organisations can be better prepared to face the challenging prospect of zero-day vulnerabilities and enhance their overall cybersecurity posture.

6.5. Data Loss Prevention Strategies

Data has become one of the most valuable assets for any organisation. It underpins business strategies, informs decision-making, and drives customer engagement. With the increasing volume and variety of data handled by organisations, coupled with the migration towards more digital platforms, the risk of data loss has significantly escalated. The repercussions of such data loss can be severe, ranging from financial penalties and reputational damage to loss of customer trust. To mitigate these risks, the implementation of robust data loss prevention (DLP) strategies is essential.

Data loss prevention refers to the set of tools and processes designed to ensure that sensitive or critical information is not lost, misused, or accessed by unauthorised users. It encompasses various strategies that can detect potential data breaches or exfiltration transmissions and prevent them by monitoring, detecting, and blocking sensitive data while in use, in motion, and at rest.

DLP strategies are designed to safeguard various types of data – from personally identifiable information (PII) such as social security numbers and credit card details to intellectual property and business-sensitive information. However, DLP is not a one-size-fits-all solution. It necessitates a nuanced understanding of the organisation's unique data ecosystem, including the nature of the data, its sensitivity, and its journey across the organisation.

Implementing a comprehensive DLP strategy involves several steps, starting with identifying the data that needs protection. This can include everything from customer databases and employee records to proprietary research and financial information. Next, it's vital to understand how this data is used and moves throughout the organisation. This data flow analysis is essential in determining potential vulnerabilities that could be exploited.

After understanding the data landscape, an effective DLP strategy requires the right technology. DLP solutions should provide tools for data identification and tracking, policy enforcement, and incident reporting and management. They should also offer flexibility to cater to the changing needs and growth of an organisation.

An often-overlooked aspect of DLP is the human factor. Employees can often be the weakest link, either due to a lack of awareness or negligence. Therefore, a successful DLP strategy also involves regular training and awareness campaigns to ensure that employees understand the importance of data security and their role in preventing data loss.

Lastly, a DLP strategy should be continuously monitored and updated to adapt to changing business needs, technological advancements, and emerging threats. Regular audits can help ensure that the DLP measures in place are effective and are being adhered to by the employees.

In conclusion, as the significance of data continues to grow, so does the need for robust data loss prevention strategies. By understanding the organisation's data landscape and implementing a multi-faceted DLP strategy, organisations can protect their valuable data assets and maintain their competitive edge in the digital era.

6.5.1. Understanding Data Loss Risks

The digital age has undoubtedly unlocked several benefits for businesses and individuals alike. Still, it has simultaneously introduced an array of data loss risks that organisations need to comprehend thoroughly. These risks can originate from a wide variety of sources, with the potential for substantial harm, ranging from financial loss to reputational damage and regulatory penalties.

One of the principal categories of data loss risks arises from cyber threats. These could include malicious activities such as hacking, ransomware attacks, and phishing scams, all intending to access, steal, or corrupt sensitive data. With increasingly sophisticated attack methods, cyber threats pose a persistent risk to organisations worldwide.

Another crucial source of data loss risks is human error. Employees can unintentionally delete data, misconfigure settings, or fall victim to phishing scams, leading to data loss. In fact, human error has been identified as one of the leading causes of data breaches in several studies, underlining the need for thorough training and awareness programs.

Software and hardware failures are another prominent source of data loss risks. Hard drives, servers, and other hardware can fail due to various reasons, including mechanical issues, power surges, or natural disasters. Similarly, software glitches or crashes can lead to sudden and significant data loss.

A less apparent but increasingly prevalent data loss risk is associated with third-party vendors or service providers. As organisations increasingly rely on external partners for various services, they inadvertently expose their data to additional risks. If these third-party providers lack robust security measures, they can become a weak link in the organisation's data security.

Moreover, data loss risks are also related to mobile and remote working. With an increase in remote work scenarios, particularly due to the recent pandemic, data is often accessed and shared over potentially insecure networks, increasing the risk of data loss. Lost or stolen devices, like laptops or smartphones, can also result in data breaches if they contain sensitive information and are not adequately secured.

Furthermore, data loss risks can stem from inadequate or outdated security measures. If an organisation's cybersecurity policies, software, or hardware are not updated regularly, they can become vulnerable to new types of threats. Similarly, non-compliance with regulations such as GDPR, CCPA, and HIPAA, which require certain data protection standards, can lead to data loss and consequent penalties.

Finally, the risk of data loss is amplified by the increasingly complex and voluminous data environments within organisations. With the proliferation of Big Data, cloud storage, and IoT devices, securing data effectively across all these varied and vast channels is a monumental challenge.

Understanding these data loss risks is the first crucial step in preventing data breaches. By identifying potential vulnerabilities and weak links, organisations can proactively implement measures to minimise these risks. This could include investing in robust cybersecurity infrastructure, providing regular employee training, enforcing stringent policies, and conducting regular audits and risk assessments.

In essence, data loss risks are multifaceted and constantly evolving. Yet, by comprehending these risks in depth, organisations can protect their valuable data assets, ensuring their sustainability and success in the digital age.

6.5.2. Implementing a Data Loss Prevention (DLP) Program

The crux of a potent cybersecurity strategy lies in its ability to prevent data loss effectively, and that's where a well-implemented Data Loss Prevention (DLP) program comes into play. A DLP program is designed to detect potential data breach incidents in real-time, prevent them, and report them to the administrators. It is a pivotal aspect of any organisation's cybersecurity arsenal.

Initiating a DLP program starts with understanding your data. This entails recognising the kind of data the organisation holds, its location, its movement within the system, and the potential risks associated with it. The data could be Personally Identifiable Information (PII), Protected Health Information (PHI), Intellectual Property (IP), financial information, or anything that, if lost, could have adverse consequences for the organisation.

The next step is to classify this data. Not all data is created equal, and hence it should not be treated equally. Certain data types may require more stringent protection measures than others. By classifying data based on its sensitivity or value to the organisation, you can ensure that appropriate protection mechanisms are in place for each data category.

Once you've understood and classified your data, the next stage in implementing a DLP program is defining the policies. These policies act as the foundation for the DLP program, dictating what constitutes a violation and the course of action if such a violation occurs. These policies should be in line with the organisation's overall risk management strategy and should comply with regulatory requirements.

With policies in place, the focus should now shift towards deploying the DLP technology. This technology should support the defined policies and provide capabilities such as data discovery, data monitoring, and incident remediation. While selecting DLP solutions, organisations should consider factors such as integration capabilities with existing systems, ease of use, scalability, and vendor support.

Furthermore, the DLP program should be equipped with comprehensive reporting and auditing capabilities. These features enable you to monitor the effectiveness of your DLP program, identify areas of improvement, and demonstrate compliance to regulatory bodies. The system should automatically generate reports on detected incidents, their severity, and the remediation actions taken.

Training is an indispensable part of implementing a DLP program. All employees should be trained on the importance of data security, the rules and regulations defined in the DLP policies, and their role in preventing data loss. This training should be a continuous process, given the ever-evolving nature of cybersecurity threats.

Lastly, a successful DLP program requires continuous monitoring and refinement. As the business environment changes, so does the nature of the data and the associated risks. Thus, the DLP program must continually evolve to cater to these changes. Regularly reviewing and updating the policies, technology, and training programs is essential to ensure the continued effectiveness of the DLP program.

Implementing a DLP program might seem like a daunting task, but its benefits far outweigh the initial efforts and investment. A robust DLP program not only protects your valuable data from unauthorised access and loss but also boosts your organisation's reputation, builds customer trust, and ensures regulatory compliance. After all, in an age where data is considered the 'new oil', its protection should be a priority for every organisation.

6.5.3. DLP Tools and Technologies: An Overview

In the context of escalating cybersecurity threats and a drastically transforming data landscape, Data Loss Prevention (DLP) tools and technologies have become an indispensable part of cybersecurity strategy. Their core objective is to safeguard sensitive data from unauthorised access, inadvertent disclosure, or outright theft.

At the heart of DLP lies the technology designed to identify, monitor, and protect data in use, data at rest, and data in motion through deep content

222

analysis. This triad covers data actively being used in various applications, data stored in databases and file servers, and data traversing networks, respectively. DLP tools offer unique features to ensure the safety of data across all these stages.

Firstly, DLP solutions possess data identification capabilities. They use techniques like pattern matching, fingerprinting, and machine learning to identify sensitive data. Pattern matching is useful for spotting structured data like credit card numbers or social security numbers, while fingerprinting and machine learning are effective for unstructured data like intellectual property.

DLP tools also offer data classification capabilities. After identifying sensitive data, these tools help classify the data based on predefined categories such as confidential, public, or internal. This classification assists in enforcing appropriate security controls based on the sensitivity of the data.

Real-time monitoring is another crucial feature of DLP tools. They monitor data in use, at rest, and in motion, providing visibility into how data is being used, who is using it, and if it's being used appropriately. If the DLP tool detects any activity that violates predefined policies, it can take corrective action, which could range from alerting the user or administrator to blocking the transaction completely.

Data encryption is also facilitated by some advanced DLP solutions. If sensitive data needs to be transmitted over unsecured networks, DLP tools can encrypt the data to prevent interception. They can also enforce the use of secure communication protocols for data transmission.

Furthermore, DLP tools aid in regulatory compliance. Many industries are bound by data protection regulations such as GDPR, HIPAA, PCI-DSS, and more. DLP tools can enforce policies aligned with these regulations, helping organisations demonstrate their compliance during audits.

Lastly, DLP solutions come equipped with robust reporting and incident management features. They can generate detailed reports on policy violations, including who violated the policy, what data was involved, and when and where the violation occurred. These reports help in conducting investigations and making informed decisions on policy adjustments.

Selecting the right DLP tool involves careful consideration of various factors. This includes understanding the nature and sensitivity of the data you handle, the regulatory landscape of your industry, your budget, and the tool's integration capabilities with existing systems.

Implementing DLP tools and technologies is a strategic move towards mitigating the risk of data loss. These solutions, while not entirely fool proof, significantly enhance an organisation's ability to protect its most valuable asset - its data. They create a protective shield around sensitive data, thereby fostering a secure data environment that can resist the evolving threats of the digital world.

CHAPTER 7. APPLICATION SECURITY

As we embark on a journey into the seventh chapter of our exploration into cybersecurity, we direct our focus towards a critical component of the digital infrastructure: application security. In an age where software applications drive business operations and personal routines, ensuring their security has never been more important.

Application security involves measures and procedures taken at the application level to prevent a myriad of potential threats that can result in unauthorised access, data theft, and service disruptions. As we move further into the age of digital transformation, the complexity and prevalence of software applications have skyrocketed, dramatically expanding the attack surface for threat actors. This is even more pronounced with the proliferation of cloud-based services, mobile applications, and IoT devices, each bringing its own set of unique security challenges.

Applications, whether they are used to control a home's heating system, process financial transactions, or store personal health records, handle a wealth of sensitive data that is attractive to cybercriminals. These applications, if not properly secured, can serve as an entry point for attackers to infiltrate an organisation's network and gain access to sensitive data.

But the scope of application security extends beyond merely protecting data. It's also about ensuring the availability and reliability of services. When applications are taken offline as a result of a security incident, businesses can suffer considerable financial losses and reputation damage.

In this chapter, we will take a deep dive into the realm of application security. We will cover topics such as common vulnerabilities in web and mobile applications, and the best practices for developing secure software. We will examine the principles of secure coding, the role of security in DevOps (often referred to as DevSecOps), and the importance of regular patching and updates.

Furthermore, we will explore the tools and technologies that can be employed to enhance application security, from static and dynamic code analysis tools to web application firewalls and vulnerability scanners. We will also delve into the standards and frameworks that guide the implementation of application security, such as the OWASP (Open Web Application Security Project) Top 10 list of the most critical web application security risks.

Ensuring the security of applications is a complex task that involves a combination of robust coding practices, comprehensive testing, timely patching, and the use of advanced security tools. Through the exploration of application security in this chapter, readers will gain a better understanding of the principles, strategies, and tools that can help protect applications from ever-evolving cyber threats. Whether you are a software developer, a cybersecurity professional, or a concerned user, this chapter aims to equip you with the knowledge needed to contribute to a safer application landscape.

7.1. Secure Coding Practices

The quality of code that builds our software applications plays a crucial role in defining their security posture. The importance of secure coding practices is more evident now than ever. They form the first and arguably the most critical line of defence against the cyber threats faced by applications.

Secure coding practices encompass guidelines and techniques used by developers to guard software applications against security vulnerabilities. The purpose of secure coding is to prevent security flaws at the source code level that could otherwise expose an application to cyber-attacks. By following secure coding practices, developers can proactively reduce the number of vulnerabilities in an application, thereby strengthening its overall security.

Understanding the potential vulnerabilities and threats that can compromise an application's security is a prerequisite to secure coding. Attackers exploit coding errors and design weaknesses in software to infiltrate systems, steal data, or disrupt operations. Examples of common coding errors include improper input validation, inadequate error handling, and weak session management, all of which can lead to critical vulnerabilities such as SQL Injection, Cross-Site Scripting (XSS), or Cross-Site Request Forgery (CSRF).

In this section, we delve deeper into the concept of secure coding, elucidating the common security vulnerabilities and exploring various coding practices that can prevent such vulnerabilities. We will cover topics such as input validation, output encoding, and memory management.

Additionally, we'll delve into some crucial principles of secure coding such as the principle of least privilege and defence in depth.

Moreover, we will address the role of secure coding in different programming languages. Every language, be it Java, Python, or C++, has its unique set of security challenges and best practices. Understanding these nuances is key to effective secure coding.

We will also discuss how to incorporate secure coding practices into different stages of the software development lifecycle (SDLC). The concept of 'shift left' in security, integrating security practices early in the SDLC, will be explored in detail. We'll examine how secure coding fits into modern development methodologies such as Agile and DevOps.

Furthermore, we will also introduce various tools and technologies that aid in secure coding, from static and dynamic analysis tools to automated code review solutions. We will also discuss the importance of developer training and awareness in secure coding, as even the most comprehensive set of guidelines and tools is ineffective without properly educated developers.

Secure coding practices, although challenging to implement and often overlooked due to pressures of rapid development and deployment, are indispensable to application security. By adhering to these practices, organisations can significantly reduce the attack surface of their applications, thereby mitigating the risk of cyberattacks. This section aims to equip developers and security professionals with the knowledge and skills required to code securely, contributing to the development of safer software applications.

7.1.1. Input Validation and Sanitisation

In software applications, user input is the driving force behind a broad range of interactions, making it both an integral component and a potential point of weakness within a system. The ability of a user to introduce data into an application provides a potential avenue for malicious actors to compromise the security and functionality of that system. This vulnerability underscores the importance of input validation and sanitisation, two practices at the heart of secure coding principles.

Input validation refers to the procedure of checking and verifying that the data supplied by a user aligns with specific rules or standards before it's processed by the system. This preventative measure aims to thwart harmful or incorrectly formed data from entering an information system. It is an effective security strategy against a myriad of security vulnerabilities, such as code injection, buffer overflow, and directory traversal attacks.

Conversely, input sanitisation goes beyond merely validating input to cleansing and modifying any data that could potentially lead to harm. Where validation may reject non-conforming data, sanitisation transforms data to render it harmless, ensuring that any damaging input is neutralised before inflicting harm.

This section will explore the importance of input validation and sanitisation, starting with an examination of the typical threats that these practices can prevent. We will look into notable instances of cyber-attacks that resulted from inadequate input validation or sanitisation to underscore the crucial role these practices play.

We will discuss a variety of ways to implement input validation and sanitisation within code. Techniques such as whitelist validation, length checks, type checks, and format checks will be examined, as well as their respective strengths and weaknesses, providing developers with practical guidance on how and when to apply each approach.

Additionally, the role of regular expressions in input validation and sanitisation will be discussed, supplemented by practical examples of how they can be employed. The importance of server-side validation, despite the prevalence of client-side checks, will be highlighted to draw attention to the dangers of over-reliance on client-side validation.

The discussion will also highlight different programming constructs and libraries that can aid in input validation and sanitisation, such as SQL's prepared statements that help prevent SQL injection attacks. We will also discuss various tools and frameworks that can automate and simplify these practices, thereby enhancing security while reducing the time and effort required in development.

Upon completion of this section, readers will have a thorough understanding of the pivotal role of input validation and sanitisation in application security. They will have practical knowledge to deploy these secure coding practices effectively, ensuring user input serves to enhance application interaction rather than become a vector for security threats.

7.1.1.1. Understanding Input Vulnerabilities

User input vulnerabilities represent a significant portion of the overall security vulnerabilities in software applications, often leading to some of the most severe consequences when exploited. Understanding the nature of these vulnerabilities is essential for establishing robust protective measures.

Input vulnerabilities can be broadly divided into two categories: Injection vulnerabilities and Inclusion vulnerabilities. Injection vulnerabilities occur when user-provided data is incorporated into a command or query that the system then executes. If the input isn't correctly sanitised or validated, a malicious user can introduce commands that manipulate the intended behaviour, leading to unauthorised access or data loss.

SQL injection is a prime example of an injection vulnerability, where an attacker manipulates user input fields to execute arbitrary SQL commands on the database, often leading to unauthorised data access or alteration. Similarly, command injection and LDAP injection also fall under this category, each posing a unique set of threats depending on the context of the application.

On the other hand, Inclusion vulnerabilities occur when user input is used to construct a path to a file or resource without proper validation. This can lead to two significant types of attacks: File inclusion attacks and directory traversal attacks. File inclusion attacks take place when an attacker can influence which file is loaded and executed by the application, often leading to remote code execution. Directory traversal attacks occur when an attacker manipulates user input to access files or directories that should be out of reach, potentially leading to unauthorised data access or disclosure.

Other user input vulnerabilities include Cross-Site Scripting (XSS), where an attacker injects malicious scripts into webpages viewed by other users, and Cross-Site Request Forgery (CSRF), where an attacker tricks a victim into performing actions on their behalf.

Understanding these vulnerabilities is crucial in devising strategies to protect applications from them. A solid foundation of secure coding practices such as input validation and sanitisation, using prepared statements or parameterised queries, proper use of access controls, and output encoding can greatly mitigate the risks associated with user input vulnerabilities. Furthermore, regular security testing, such as fuzz testing and penetration testing, can help identify any potential input vulnerabilities in an application, facilitating early detection and remediation.

7.1.1.2. Implementing Effective Input Validation Techniques

User input validation is one of the most fundamental, yet often overlooked, aspects of application security. It's the first line of defence against a wide array of attack vectors that could exploit poorly sanitised or unverified input. Implementing effective input validation techniques can significantly improve an application's resilience to these attacks.

At its core, input validation involves checking the user's input against a set of predetermined criteria or rules before processing it. This could be as simple as ensuring that a zip code consists of five digits, or as complex as verifying the syntax and semantics of a piece of code in a web form. The primary goal is to ensure that only safe and appropriate data is processed by the system, mitigating the risk of injection attacks and other common exploits.

There are several techniques to effectively implement input validation. The first step is to identify where in the application user input is accepted, and what form this input should take. Knowing the expected type, length, format, and range of acceptable values is crucial.

One effective technique is to use a whitelist approach, where only input that meets specific criteria is accepted. For example, an application might only accept alphanumeric characters in a username field, rejecting any input that includes special characters. A whitelist approach is typically more secure than a blacklist approach, which seeks to block known harmful input, as it's easier to define what is acceptable than to predict all potential harmful input.

Another useful technique is to use built-in validation functions or libraries. Most programming languages and frameworks provide functions for checking the format and type of data, which can be used to validate user input. For example, regular expressions can be used to check if a string matches a specific pattern.

Input sanitisation is another important aspect of input validation. It involves cleaning up the input by removing or escaping special characters that could be used in an attack. It's particularly important when dealing with input that will be included in a command or query or displayed back to the user.

For example, if user input is inserted into a SQL query, it's important to escape special characters to prevent SQL injection attacks. This can be done using functions provided by the database management system, or by using parameterised queries or prepared statements, which automatically handle this escaping.

It's also important to consider where validation should occur. Client-side validation can improve user experience by providing immediate feedback, but it should never be relied upon for security, as it can be easily bypassed. Therefore, server-side validation is essential to ensure security.

Implementing effective input validation can be complex, particularly for large applications, but it's a critical component of secure coding. Regular code reviews and security testing can help ensure that validation is correctly implemented and effective in protecting against attacks.

7.1.2. Principle of Least Privilege in Application Development

The principle of least privilege (PoLP) is a foundational concept in security that can and should be applied to various aspects of an organisation's operations, including application development. Essentially, it mandates that any process, program, or user should have only the bare minimum privileges necessary to perform its tasks and no more. This reduces the risk of unauthorised access or misuse of information.

In the context of application development, adopting the principle of least privilege involves designing and coding applications in such a way that they request and operate with the fewest system privileges possible. From the perspective of a developer, this concept may seem counter intuitive as it might be more convenient to work with higher levels of access or privileges. However, it's crucial for the security posture of the application, especially when mitigating the impact of potential software vulnerabilities.

In practice, applying PoLP in application development takes a variety of forms. One common example is running applications with non-administrative privileges, even when executed by an administrative user. This reduces the risk of an application being exploited to carry out privileged operations. Similarly, application components and subsystems should also operate under least privilege, with access restrictions clearly defined for each component or subsystem.

Another critical application of PoLP is in the area of database access. Applications should use distinct accounts for connecting to databases, with each account having the minimum necessary access rights. For instance, an application that only needs to read data from a specific database should not have permissions to write to that database.

Furthermore, adopting PoLP can extend to code libraries and third-party components used within the application. Use only those libraries that are necessary and grant them the least privilege necessary for the application to function properly.

Applying the principle of least privilege also involves ongoing review and updates. As applications evolve, so do their privilege requirements. Regularly reviewing these privileges can ensure they remain at the minimum necessary level. For example, if a feature that required special permissions is deprecated, the associated privileges should be revoked.

In conclusion, integrating the principle of least privilege into application development minimises potential attack vectors and the possible damage from a successful exploit. It's a key best practice for creating secure, resilient applications. Embracing this principle requires a mindset that values security, and an ongoing commitment to review and reduce privileges wherever possible.

7.1.2.1. Implementing Least Privilege in Code

Implementing the Principle of Least Privilege (PoLP) in code requires careful planning and thorough consideration from the earliest stages of application design and development. It's a best practice that extends to all aspects of software creation, affecting choices at the level of architecture, programming language, libraries, data access, and more. Here are several key steps and considerations when applying PoLP to your code.

1. Design and Architecture: At the very outset, design your software architecture with least privilege in mind. This could include isolating different functionalities into separate modules or services, each running with only the privileges they need to perform their designated tasks.

2. Use of Roles and User Accounts: If your application includes user accounts, assign roles to limit what each account can do. This limits the harm that could be done if a user account is compromised. Likewise, ensure the application uses unique identifiers, rather than generic or shared accounts, to access other system resources.

3. Restricting File and Network Access: Software often requires access to files and network services. These privileges should be as narrow as possible. For example, an application may need read access to a particular directory, but not write access. Likewise, it might need to make outgoing network connections, but not accept incoming connections.

4. Handling of Sensitive Data: If an application must handle sensitive data, such as passwords or personal identifiable information (PII), ensure the code handling this data has minimal privileges. Avoid storing sensitive data in areas with broad access.

5. Database Access: Many applications need to access a database. Ensure that they use a specific database account with minimal rights, rather than a high-level administrative account. Use prepared statements and parameterised queries to avoid SQL injection attacks.

6. Error Handling: Code often has more privileges than it needs because of poor error handling. By correctly managing exceptions and implementing good error handling, you can avoid situations where code needs more privileges to recover from errors.

235

7. Libraries and Dependencies: When using third-party libraries or other dependencies, follow the PoLP as well. If the library only needs to perform a specific task, there's no need for it to have broad system access.

8. Regular Code Reviews and Audits: To ensure the principle of least privilege is consistently applied, regular code reviews and audits are essential. Automated tools can help identify privilege escalations, but human review is invaluable for maintaining good security hygiene.

Implementing the principle of least privilege in code is an ongoing process that should be an integral part of your software development lifecycle. By committing to this practice, developers can significantly enhance the security of their applications, making them more resistant to exploitation even if vulnerabilities are present.

7.1.2.2. Balancing Security and Functionality

Implementing the Principle of Least Privilege (PoLP) in software development inherently requires a delicate balancing act between ensuring security and maintaining functionality. When well-executed, this approach offers robust security without compromising the utility or performance of the application. Here are key considerations and strategies to balance security and functionality when adhering to PoLP in application development:

1. User Experience (UX) Considerations: User experience is a pivotal aspect of any application. Developers need to implement security measures, such as role-based access controls or multi-factor authentication, without disrupting user workflows or productivity. UX designers and developers should collaborate closely to ensure security enhancements do not negatively impact user satisfaction or hinder functionality.

2. Gradual Permission Escalation: To ensure a smooth user experience while maintaining security, consider implementing gradual permission escalation in your applications. This means users start with minimum permissions and privileges are only escalated when necessary for a specific task, often with additional user verification. This strategy keeps the application secure most of the time and only elevates privileges when absolutely needed.

3. Modularity and Microservices: A modular architecture or microservices-based design can help balance security and functionality. By breaking the application down into smaller, isolated components, each part can run with only the necessary privileges, reducing the potential attack surface. This design also promotes scalability and easy updates without impacting the entire system.

4. Thorough Testing: Conduct comprehensive security testing alongside functionality testing. Tools such as static and dynamic application security testing (SAST and DAST) can help identify security flaws without compromising functionality. Penetration testing and red teaming exercises can also validate that security measures are effective and do not interfere with application operations.

5. Continuous Monitoring and Updates: Post-deployment, continue to monitor the application's behaviour and its interaction with system resources. Regular updates, patching, and adjustments may be necessary to address newly discovered vulnerabilities or to tweak privilege levels to better balance functionality and security.

6. Security Training for Developers: Provide regular security training to developers. Understanding secure coding principles, such as PoLP, helps them naturally produce code that is secure yet functional. Also, a security-aware development team is less likely to introduce security flaws in the name of functionality.

7. Automation: Automate security controls where possible. Automation can limit human error, a common source of security vulnerabilities, and can enforce security measures without adding friction to the user experience.

In conclusion, while balancing security and functionality in application development can be challenging, it is essential for creating applications that are both useful and secure. By incorporating secure design principles and strategies, developers can minimise risk while ensuring their applications fulfil their intended functions and deliver a positive user experience.

7.2. Web Application Security

As the internet continues to be an integral part of our daily lives, web applications have evolved into sophisticated platforms that serve various purposes, from online banking and shopping to social media and remote work tools. This widespread use of web applications makes them a lucrative target for cybercriminals, necessitating a robust framework for web application security.

Web application security is a critical component of cybersecurity that focuses on protecting websites and online services against different types of threats and attacks, including but not limited to SQL injection, cross-site scripting (XSS), and cross-site request forgery (CSRF). It is essential not only for the protection of sensitive data these web applications may hold but also for maintaining users' trust in digital platforms.

Web applications often serve as the public face of an organisation. Hence, any successful cyber-attack could result in severe reputational damage and financial loss, not to mention regulatory fines in case of a data breach. The security of these applications is, therefore, a top priority for organisations.

Web application security is an interdisciplinary field, encompassing multiple areas like secure coding practices, use of security headers, SSL/TLS encryption, application firewall configurations, and regular vulnerability scanning and testing among others. A secure web application is one where these security measures are implemented in all stages of development and operation, from design and development to deployment and maintenance.

In addition, security practices should include continual training for developers, emphasising secure coding techniques, understanding of common web application vulnerabilities as described by the OWASP Top 10, and proactive engagement in threat modelling exercises. The latter encourages developers to think like an attacker, helping to identify and patch potential security flaws before they can be exploited.

The fast-evolving landscape of web application security calls for a proactive, rather than reactive, approach. With new vulnerabilities discovered frequently, organisations must remain vigilant and up to date on the latest threats and security practices. This includes regular updates and patches to their web applications, a process that should be made easier through the use of automatic updates or CI/CD pipelines.

Looking ahead, advancements in technologies such as machine learning and AI offer promising avenues for enhancing web application security. These tools can aid in automatically detecting anomalous behaviour or identifying potential vulnerabilities more quickly and accurately. However, they should be seen as a complement to, not a replacement for, traditional security practices.

In summary, web application security is a complex, ever-evolving discipline, but it is a non-negotiable aspect of modern digital operations. As we dive deeper into this chapter, we will explore various facets of web application security, including specific types of threats, mitigation strategies, and best practices to build and maintain secure web applications.

7.2.1. Common Web Application Vulnerabilities

Web application vulnerabilities are software flaws or misconfigurations in a web application that an attacker can exploit to gain unauthorised access, perform unauthorised actions, or expose sensitive data. These vulnerabilities could lead to severe consequences, including data breaches, financial loss, and damage to an organisation's reputation.

One of the most prevalent web application vulnerabilities is injection flaws, such as SQL, OS, and LDAP injection. Injection flaws occur when an attacker can send malicious data to an interpreter through a web application. For instance, SQL injection involves an attacker manipulating a query to the database that can lead to unauthorised data access or manipulation.

Cross-Site Scripting (XSS) is another common web application vulnerability, which allows an attacker to inject malicious scripts into web pages viewed by other users. This could lead to session hijacking, identity theft, defacement of websites, or even distribution of malware.

Similarly, Cross-Site Request Forgery (CSRF) tricks a victim into submitting a malicious request, typically targeting state-changing requests, not data theft, since the attacker has no way to see the response to the forged request. An attacker successfully exploiting a CSRF vulnerability could force a user to perform actions they didn't intend to, such as changing their email address or password.

Insecure Direct Object References (IDOR) is another issue, where a web application exposes a reference to an internal implementation object. An attacker can manipulate these references to access unauthorised data.

Misconfiguration of security settings is a frequent issue that can lead to unauthorised access to sensitive information or functionality. This could be something as simple as leaving default credentials in place or failing to properly set security headers.

Other common vulnerabilities include issues related to session management, unvalidated redirects and forwards, and insecure use of cryptography. Furthermore, newer classes of vulnerabilities are emerging with the increased use of APIs, cloud based environments, and advanced JavaScript frameworks in web application development.

Understanding these common vulnerabilities is just the first step towards building secure web applications. Equally critical is adopting strategies to mitigate these vulnerabilities, including secure coding practices, thorough testing, and keeping up to date with the latest security updates and patches. Furthermore, developers must have a deep understanding of the HTTP protocol and the behaviour of web browsers, as well as the security implications of their code.

As we proceed in this chapter, we will delve deeper into these vulnerabilities, their potential impact, and the best practices and strategies for mitigating these risks. Through this knowledge, organisations can better protect their web applications from attacks, thereby safeguarding their data and maintaining trust with their users.

7.2.1.1. Cross-Site Scripting (XSS) and SQL Injection Attacks

Two types of attacks that frequently plague web applications are Cross-Site Scripting (XSS) and SQL Injection. Both of these attacks exploit vulnerabilities in a web application to perform unauthorised actions, and both can have serious security implications.

Cross-Site Scripting (XSS) is a type of injection attack in which malicious scripts are inserted into trusted websites. Unlike other web attacks, XSS directly targets the user rather than the application itself. This attack can occur when a web application allows user input to be included without

241

proper validation or escaping, and this user input is then rendered on the webpage.

There are three primary types of XSS attacks: stored, reflected, and DOM-based. Stored XSS attacks occur when malicious scripts are permanently stored on the target server. In reflected XSS attacks, the malicious script is embedded in a URL, which then reflects the script off the web server to the user's browser. DOM-based XSS attacks occur entirely in the victim's browser where the original client-side script manipulates the Document Object Model (DOM) environment in the victim's browser.

The consequences of an XSS attack can be severe. If an attacker manages to inject a malicious script into a webpage that a user, then executes, the attacker can gain the ability to impersonate or masquerade as the victim user, carry out any actions that the user is able to perform, and gain access to any information the user is able to access.

SQL Injection, on the other hand, is an attack technique where an attacker inserts malicious SQL code into a database query. The primary form of defence against SQL injection is input validation and parameterised queries, including prepared statements. The principle behind these defence strategies is that all data from external sources - including user inputs - are inherently untrusted and must be validated before use.

The danger of SQL injection attacks is their potential to manipulate the commands that a web application sends to a database, allowing an attacker to view, alter, or delete data that they typically wouldn't have authorisation to access. In some cases, SQL injection can even be used to execute commands on the operating system, potentially allowing an attacker to escalate to more privileged accounts.

Both XSS and SQL injection attacks represent significant threats to web application security, and both can be catastrophic if not properly guarded against. Implementing comprehensive defence strategies and educating developers about these vulnerabilities is an essential part of any robust web application security program. Furthermore, regular security audits and penetration testing are also crucial to identify and fix any potential vulnerabilities before they can be exploited.

While securing a web application against these threats can be challenging, the importance of such measures cannot be overstated. The financial and reputational damage resulting from a successful XSS or SQL injection attack can be enormous, emphasising the need for proper security mechanisms to mitigate these risks. As we explore further in this chapter, we will discuss more in-depth methods of securing applications against these and other threats, ensuring a robust line of defence against potential attackers.

7.2.1.2. *Mitigating Common Web Application Vulnerabilities*

Mitigating common web application vulnerabilities is a multi-faceted task that requires both technical prowess and a comprehensive understanding of various threat vectors. The goal of mitigation strategies is not just to patch existing vulnerabilities, but to build resilient systems that can withstand and recover from attacks.

When it comes to mitigating Cross-Site Scripting (XSS) and SQL Injection vulnerabilities, the two primary strategies revolve around input validation and output encoding.

Input validation involves verifying that all user-supplied input is expected and conforms to predetermined formats. This includes checking for length, type, and syntax, among other factors. This practice is fundamental for preventing both XSS and SQL Injection attacks, as it ensures that potentially harmful inputs are either rejected or sanitised before they can do any harm. For example, web applications should ideally be designed to reject all HTML and script tags from user input fields to mitigate against potential XSS attacks.

Output encoding, or escaping, involves treating received data as displayable text, rather than executable code. This involves encoding special characters in a way that the browser will display them as they are, instead of executing them as part of a script. For instance, output encoding would transform a script tag into a string that the browser would render as visible text, rather than a potentially harmful script.

However, input validation and output encoding alone are not sufficient for mitigating these common web vulnerabilities. Other best practices include:

- Least Privilege Access: Restricting the permissions of database accounts used by web applications can limit the potential damage caused by a successful SQL injection attack.

- Use of Prepared Statements: By using prepared statements or parameterised queries, SQL queries can be constructed safely, rendering SQL injection attacks ineffective.

- Content Security Policy (CSP): This is a security layer that helps detect and mitigate certain types of attacks, including XSS and data injection attacks. A CSP can be used to specify the domains that a browser should consider as valid sources of executable scripts, effectively preventing the execution of malicious scripts injected by an attacker.

- Regular Security Audits: Regularly scanning and testing web applications for vulnerabilities is crucial. This includes manual code reviews, static analysis tools, and dynamic analysis tools such as fuzzing.

- Educating Developers: The root cause of many vulnerabilities is often a lack of awareness or understanding. Providing developers with the necessary training to understand and avoid security pitfalls is a proactive measure that can significantly reduce the risk of introducing vulnerabilities into web applications.

Through these methods and a vigilant approach to security, it's possible to mitigate the common web application vulnerabilities like XSS and SQL injection, thereby increasing the robustness of web applications and the safety of their users. This strategic, multi-layered approach to security forms the foundation of effective web application protection.

7.2.2. Securing APIs

Securing Application Programming Interfaces (APIs) is an integral part of web application security. APIs serve as the bridge between different software systems, allowing them to interact and exchange data. Given the sensitive nature of the data that often passes through APIs, ensuring their security is paramount.

API security involves implementing measures that ensure only authorised requests are processed and all exchanged data is adequately protected. Securing APIs can be a complex task due to their unique nature and the varied threats they face. However, following certain guidelines can significantly enhance the security posture of APIs.

Authentication and authorisation play a vital role in API security. Authentication is about confirming the identity of the requesting party, while authorisation ensures they have the necessary permissions to perform the requested action. Token-based authentication, such as OAuth 2.0 or JWT (JSON Web Tokens), is often used due to its scalability and security advantages.

Rate limiting is another essential part of API security. By limiting how many requests a client can make within a specified timeframe, you can mitigate the risk of Distributed Denial of Service (DDoS) attacks and prevent resource exhaustion.

Input validation is as important for APIs as it is for web applications. APIs should validate all incoming data to ensure it meets expected criteria and doesn't contain malicious content. This measure helps prevent attacks such as SQL injection and Cross-Site Scripting

Encrypting data in transit and at rest is crucial for maintaining the confidentiality and integrity of the data. Using HTTPS for data transmission is a standard practice, but it's also important to consider data encryption at rest, particularly for sensitive information.

API versioning is another strategy to enhance security. When updates or changes are made to the API, these can be released as new versions. This allows deprecated versions to be phased out, reducing the risk of attackers exploiting outdated, less secure versions.

Monitoring and logging are key to maintaining visibility into the API's operations and identifying potential security issues. Monitoring can help detect anomalies that may indicate an attack, while logging provides a record of events for forensic analysis in case of a security incident.

Incorporating security testing into the development lifecycle of the API can help identify and resolve potential security vulnerabilities early. This can include both static and dynamic security testing methods.

Finally, a robust error handling mechanism is essential. In case of an error, the API should return generic error messages to avoid leaking sensitive information that could be exploited by an attacker.

By adopting these practices, organisations can improve the security of their APIs, thereby protecting the integrity and confidentiality of the data they handle and enhancing the overall security of their web applications.

7.2.2.1. Understanding API Security Risks

APIs, or Application Programming Interfaces, form the backbone of modern software connectivity and integration. As crucial as they are, APIs also introduce a range of security risks that could potentially lead to data breaches and system compromise. A thorough understanding of these risks is essential in designing, developing, and implementing secure APIs.

API security risks often stem from misconfigurations, design flaws, or inadequate security measures. Here, we delve into several common API security risks and provide insights on their implications.

Inadequate Authentication/Authorisation: If APIs fail to authenticate and authorise users effectively, it could lead to unauthorised data access or manipulation. In cases where API keys are used, they may be improperly stored or transmitted, leading to key leakage. In the worst case, this could lead to full system compromise.

Excessive Data Exposure: APIs often return more data than required by the client. Attackers can take advantage of this excessive data exposure to obtain sensitive information that they should not have access to.

Unsecured Communications: If APIs communicate data over unencrypted channels, sensitive data could be intercepted during transmission. This could include personal user data, API keys, or other confidential information.

Injection Attacks: APIs that do not adequately sanitise input data may fall victim to injection attacks. For example, an attacker could manipulate API requests to perform actions not intended by the application developers, such as SQL Injection, or even run malicious scripts.

Poor Error Handling: Improperly handled errors can leak sensitive information, providing attackers with valuable insights into the system's inner workings. For example, detailed error messages could reveal system vulnerabilities or valuable system information that could aid an attack.

Broken Object Level Authorisation (BOLA): BOLA occurs when APIs expose endpoints that handle object identifiers, creating a possibility for attackers to manipulate these IDs to gain unauthorised access to resources.

Lack of Rate Limiting: Without proper rate limiting, APIs are vulnerable to brute force attacks and Denial of Service (DoS) attacks. Attackers could overload the API with requests, leading to system slowdowns or crashes.

Inadequate Monitoring and Logging: Without proper monitoring and logging, malicious activity might go unnoticed until it's too late. Detailed logs can provide critical information during incident response and post-mortem analysis.

API Versioning Issues: If older, less secure versions of APIs are not deprecated correctly, they can serve as an entry point for attackers. Ensuring the secure retirement of outdated API versions is as important as securing current ones.

Understanding these risks is the first step in securing APIs. Once these risks are identified, mitigating strategies involving strong authentication and authorisation, thorough input validation, robust error handling, secure communication, and comprehensive monitoring and logging can be employed to bolster API security.

7.2.2.2. Best Practices for Secure API Design

Ensuring API security is paramount in today's technology landscape, given the extensive usage and reliance on APIs for system integrations and functionality enhancements. API security hinges on careful design and implementation. Here, we will delve into best practices for secure API design, drawing on industry-accepted principles and strategies:

1. **Implement Strong Authentication and Authorisation:** Authentication verifies who the user is, while authorisation determines what they are allowed to do. Use protocols like OAuth 2.0 for delegated authorisation and OpenID Connect for user authentication. Consider using API keys or JWT tokens for API authentication, but ensure they're stored and transmitted securely.

2. **Enforce Principle of Least Privilege:** Ensure that API consumers have only the minimal permissions necessary to perform their tasks. This limits the potential damage in case of a security breach.

3. **Employ Robust Input Validation:** Validate and sanitise all input data to prevent injection attacks. Utilise allow-lists instead of deny-lists, specifying the exact input patterns you expect.

4. **Use HTTPS for Secure Communication:** All API communication should be done over HTTPS to ensure the data in transit is encrypted and secure from man-in-the-middle attacks.

5. **Rate Limit API Requests:** Implement rate limiting on your API endpoints to mitigate the risk of DDoS attacks and brute force attacks.

6. **Minimise Data Exposure:** Always follow the principle of providing the minimum necessary data in API responses. This practice, also known as data minimisation, can help reduce the risk of exposing sensitive information.

7. **Implement Effective Error Handling:** Avoid revealing sensitive information through error messages. Detailed internal error information should be logged and not exposed to the client.

8. **Ensure Proper Versioning:** Always version your APIs so that changes do not break the applications using your APIs. This also allows you to deprecate insecure or outdated versions in a controlled manner.

9. **Perform Regular Security Audits:** Use automated tools to conduct security audits on your APIs regularly. This will help you detect any security vulnerabilities in your API endpoints.

10. **Use a Web Application Firewall (WAF):** A WAF can help protect your APIs against common web exploits and attacks, like SQL injection and XSS.

11. **Monitor and Log API Usage:** Implement a robust logging and monitoring system to track API usage and detect any suspicious activity early. This aids in quick incident response and post-mortem analysis.

Secure API design involves a thoughtful approach to security at every stage of the API lifecycle, from design and development to deployment and monitoring. By following these best practices, organisations can significantly bolster the security of their APIs and mitigate potential cyber risks.

7.3. Database Security

The importance of robust database security can't be overstated in our data-driven world. Databases often hold sensitive and valuable information, making them prime targets for cybercriminals. Implementing comprehensive database security measures is critical to safeguard data integrity, confidentiality, and availability. In this section, we'll explore key concepts, strategies, and best practices in database security.

Databases serve as the foundation for countless applications and systems. They store a wide variety of information types, including personal user data, financial transactions, health records, and proprietary company information. The security implications are far-reaching, impacting individuals, businesses, and even nations. A breach can result in identity theft, financial fraud, regulatory penalties, and severe reputational damage.

The nature of database security is multifaceted, with numerous potential vulnerabilities. These can exist at various levels, including the physical server environment, the database management system (DBMS), the network connections, and the stored data itself. Consequently, effective database security necessitates a comprehensive, multi-layered approach.

This approach entails numerous elements, including securing the physical and virtual infrastructure hosting the database, implementing stringent user access controls, ensuring secure data transmission, regular monitoring and auditing of activities, and effective disaster recovery planning. Additionally, adhering to best practices in database design and maintenance, such as data minimisation, encryption, and timely patching of vulnerabilities, contributes significantly to overall database security.

Apart from these technical aspects, human factors play a critical role in database security. A high percentage of data breaches can be traced back to human error or insider threats. As such, fostering a culture of security awareness, enforcing strict operational protocols, and maintaining an updated understanding of evolving threats are vital in mitigating database security risks.

Looking ahead, database security continues to face new challenges and opportunities. The proliferation of big data, cloud-based databases, and increasingly sophisticated cyber threats necessitate continual innovation and vigilance in database security strategies. The future of database security will likely see increasing reliance on advanced technologies such as artificial intelligence and machine learning to enhance threat detection and response capabilities.

In the following sections, we'll delve deeper into specific aspects of database security, providing practical insights to help you bolster your defences and protect your valuable data assets. Whether you're a database administrator, an application developer, a cybersecurity professional, or an executive responsible for data governance, understanding database security is an essential part of navigating our interconnected digital world.

7.3.1. Common Database Security Threats

Database security threats are numerous and continually evolving as cybercriminals find new ways to breach defences. Given the wealth of sensitive data that databases typically contain, understanding these threats is critical. In this section, we explore some of the most common threats to database security.

SQL Injection Attacks: This remains one of the most prevalent threats to databases. Attackers manipulate SQL queries, typically via user inputs in a web application, to execute malicious commands on the database. These attacks can lead to unauthorised data access, data corruption, or even data loss.

Unauthorised Access: Unauthorised access can occur due to weak or insufficient authentication mechanisms, poorly managed user permissions, or exploitation of software vulnerabilities. This can result in data theft, manipulation, or deletion.

Insider Threats: Malicious actions or negligence by trusted insiders can pose a significant threat. These can range from employees accessing data without proper authorisation to administrators neglecting security measures or even engaging in malicious activities themselves.

Database Malware: Databases can be infected with various types of malware, including viruses, worms, or ransomware, which can corrupt data, degrade performance, or render the database inaccessible.

Data Leakage: This refers to the unintentional exposure of data due to factors like misconfigurations, weak security controls, or human error. It can lead to significant breaches of confidential data.

Denial of Service (DoS) Attacks: These attacks are designed to overwhelm a database with requests, causing it to slow down significantly or even crash, thereby denying service to legitimate users.

Exploiting Backup Vulnerabilities: Backup systems, if not properly secured, can provide an alternative route for attackers to access a database. Moreover, backups can also leak data if they're not encrypted or if they're stored or disposed of insecurely.

These threats underscore the need for comprehensive, multi-layered security measures to protect databases. These include regular patching and updates, strict access controls, robust authentication mechanisms, encryption, regular auditing and monitoring, secure backup procedures, and staff training to recognise and mitigate potential threats.

SQL injection attacks, for instance, can be mitigated by input validation and parameterised queries. Unauthorised access can be curbed with strong password policies, two-factor authentication, and least privilege access controls. Insider threats can be managed by regularly reviewing user permissions, conducting background checks, and implementing strict operational protocols.

To combat database malware, it's critical to keep the DBMS updated, use reputable antivirus software, and maintain a robust firewall. Preventing data leakage involves measures such as configuring databases securely, encrypting sensitive data, and regularly auditing database activities. DoS attacks can be mitigated with traffic filtering, rate limiting, and implementing redundant systems to ensure availability.

Ensuring secure backups involves encrypting backup data, storing backups securely, and regularly testing restore procedures. Regular security audits can help identify potential vulnerabilities in backup processes.

In summary, while the threats to database security are numerous and evolving, a proactive and comprehensive approach to security can significantly mitigate these risks and protect valuable data assets. The next sections will delve into more details about implementing these security measures.

7.3.2. Database Encryption and Access Control

The foundational principles of database security rest on two pillars: Encryption and Access Control. Both mechanisms are crucial in creating a secure database environment, preserving data integrity, and preventing unauthorised data disclosure.

Database Encryption: In the digital world, encryption is akin to a high-strength padlock. It ensures that even if data is accessed or intercepted without authorisation, it remains unintelligible and useless to the viewer without the decryption key.

Database encryption typically involves two types: Data-at-rest encryption and Data-in-transit encryption. Data-at-rest encryption is applied to data that is stored on disk within the database. This kind of encryption can protect the data even if the physical storage medium (e.g., hard drives, SSDs) is stolen or accessed by an unauthorised individual.

On the other hand, Data-in-transit encryption safeguards the data while it is being transferred between the database and the client application or between two database servers. Techniques such as SSL/TLS are often employed to secure data in transit. Regardless of the type of encryption used, key management plays a crucial role in maintaining encryption effectiveness.

Database Access Control: Effective access control ensures that only authorised individuals have access to the database, and even within that, they can only access the data and perform the operations that they're permitted to. Access control starts with robust authentication mechanisms, such as passwords, two-factor authentication, or biometric verification.

Once authenticated, a user should be subject to authorisation rules based on the principle of least privilege (POLP), which means users should be granted only the minimum permissions necessary to carry out their tasks. This approach minimises the potential damage if an account is compromised.

Role-Based Access Control (RBAC) is commonly used to manage user permissions in a database. In RBAC, roles are created corresponding to various job functions, and permissions are assigned to these roles. Users are then assigned roles, determining their access rights. This system can simplify the management of permissions, especially in large organisations.

Furthermore, session management is another crucial aspect of access control. It involves tracking and controlling active database sessions. For example, idle sessions might be automatically logged out after a certain period, and the number of simultaneous sessions might be limited to prevent abuse.

Both database encryption and access control need to be complemented with regular security audits, intrusion detection systems, and ongoing employee training programs. It's important to note that while these measures significantly enhance security, they are not entirely fool proof, and a robust disaster recovery plan is also needed as a final line of defence.

While encryption secures the data against unauthorised viewing, access control determines who can do what within the database. Used in combination, they provide a potent defence against the majority of database security threats. In the subsequent sections, we will delve deeper into the specifics of these strategies and how they can be employed effectively.

7.3.3. Database Security Auditing and Monitoring

Maintaining the integrity of a database goes beyond merely establishing robust defence mechanisms. It is a dynamic process that requires continuous surveillance and periodic inspection. The essence of these practices is captured in the processes of Database Security Auditing and Monitoring.

Database Security Auditing: Security auditing involves the systematic review and examination of database activities and structures. It's like a health check-up for your database security measures. The goal is to identify vulnerabilities, ensure compliance with security policies, and validate the effectiveness of current controls.

An audit can cover many areas, such as examining user accounts to ensure proper privilege settings, checking password policies, reviewing database configurations, inspecting access control measures, and verifying that encryption is properly implemented. In some situations, an audit might involve testing the system for vulnerabilities using penetration testing methods.

The outcome of a security audit is typically a detailed report outlining discovered weaknesses and providing recommendations for remediation. Regular security audits are crucial because databases, their environments, and associated threats are continually evolving. A configuration that was secure yesterday may become vulnerable tomorrow due to the discovery of a new exploit or the release of a new version of database software.

Database Security Monitoring: While auditing provides a snapshot of the database security at a specific time, security monitoring is an ongoing process that keeps a watchful eye on the database environment in real-time.

Monitoring involves tracking and analysing database activity to detect and respond to potential threats. It may involve observing login activities, tracking SQL queries for signs of SQL injection attempts, monitoring changes to database structures, or checking for unusual patterns of data access. Automated tools, often part of Security Information and Event Management (SIEM) systems, are typically used to gather and analyse this data, raising alerts when suspicious activities occur.

Monitoring should also encompass performance tracking. Sudden performance changes can indicate potential security incidents. For instance, a sudden increase in CPU usage could suggest a Denial of Service (DoS) attack, and a spike in data read operations might indicate a data leak.

Furthermore, security logs play a vital role in monitoring. They provide a history of actions and can be invaluable in investigating security incidents or demonstrating compliance with regulations. Therefore, logs should be kept secure, complete, and regularly reviewed.

In conclusion, database security auditing and monitoring are complementary processes that contribute to a comprehensive security strategy. Auditing provides periodic, in-depth security inspections, while monitoring ensures real-time surveillance of database activities. Together, they create a dynamic and robust security structure that evolves with the ever-changing landscape of cybersecurity threats. In the coming sections, we will further explore how to leverage these processes effectively in different contexts.

7.4. Security Testing and Auditing for Applications

Security Testing and Auditing for Applications are critical components of any comprehensive cybersecurity strategy, acting as the primary mechanisms for validating the efficiency and effectiveness of security measures put in place. They provide the feedback loop necessary for an organisation to adapt and improve its security posture over time.

Security testing, the first of these practices, is an investigative process where applications are deliberately probed for weaknesses. It's not a one-off process, but an ongoing practice that should be ingrained in the entire software development lifecycle. It helps in identifying vulnerabilities in applications that could be exploited by attackers, including those in coding and design. Several types of security testing exist, including static and dynamic analysis, penetration testing, fuzz testing, and more. Each offers unique perspectives and reveals different potential vulnerabilities.

Moreover, security testing isn't only about finding vulnerabilities. It also gauges an application's reaction to different types of attacks. This involves understanding how the application handles malicious inputs or excessive traffic, whether it logs events correctly, how it recovers from crashes, and if any information is leaked when it fails.

On the other hand, security auditing is a more formal and systematic approach, typically performed by independent entities. An audit's purpose is to ensure that the applications and associated processes comply with the organisation's security policies, regulatory requirements, and industry best practices. The auditing process examines various aspects such as system configuration, access controls, encryption mechanisms, incident response procedures, and more.

An essential aspect of a security audit is the audit trail - detailed logs that record all actions performed. These logs serve as evidence of compliance with the defined policies and provide crucial information during forensic investigations.

In essence, while security testing is more focused on the applications themselves, security auditing takes a broader view, assessing the entire environment and procedures related to the applications. Both are necessary to achieve a secure posture.

Understanding and implementing security testing and auditing for applications is an absolute necessity in today's cybersecurity landscape, not just a 'nice-to-have' element. It is no longer a question of 'if' an attack will occur but 'when'. A robust security testing and auditing process can mean the difference between a minor security incident and a catastrophic system compromise. The following sections will delve deeper into the methodologies, techniques, and best practices involved in security testing and auditing for applications.

7.4.1. Understanding Security Testing Techniques: SAST, DAST, IAST

Security Testing Techniques play a pivotal role in identifying potential vulnerabilities and flaws in the system and applications. Three prominent types of security testing techniques are Static Application Security Testing (SAST), Dynamic Application Security Testing (DAST), and Interactive Application Security Testing (IAST). Each testing technique has its unique process, benefits, and scope of operation, contributing to a robust and comprehensive security evaluation.

SAST, often referred to as "white-box testing," involves testing the source code of an application. It's designed to analyse the code at a static stage, i.e., the code is not executed during the testing process. Instead, SAST examines the code, byte compilations, and binaries for common coding flaws and security loopholes. This is especially useful in identifying issues such as buffer overflows, SQL injections, or improper encryption techniques. SAST is typically used in the development stage as part of the Secure Software Development Life Cycle (SSDLC) as it helps developers understand and rectify the issues early in the development process, thereby saving significant time, cost, and potential risk exposure.

In contrast, DAST or "black-box testing," involves testing an application in its running state. Unlike SAST, which looks into the internal structure of the applications, DAST evaluates the application from an outsider's perspective. It uses fault injection methods to identify vulnerabilities that might be exploited by an external attacker, such as cross-site scripting (XSS) or server misconfigurations. The advantage of DAST is that it tests the application in a real-world scenario, simulating potential attacks to identify vulnerabilities that might be visible only during runtime. However, it doesn't provide the depth of analysis of the source code that SAST does.

IAST is a security testing method that combines aspects of both SAST and DAST to provide a more comprehensive view of application security. It's often referred to as "grey-box testing." IAST involves the use of agents within the application to monitor the data flow and identify the behaviour that could indicate a security vulnerability. This technique combines the advantages of SAST's code-based vulnerability identification and DAST's runtime analysis. The result is a more accurate, thorough, and faster identification of potential vulnerabilities.

A noteworthy feature of IAST is its ability to identify the exact lines of code where a vulnerability exists during the application's execution. This precision makes it much easier and quicker for developers to remediate identified vulnerabilities. Additionally, because IAST works during the application's run-time, it's less likely to report false positives, a common problem with both SAST and DAST.

Each of these techniques—SAST, DAST, and IAST—plays an essential role in the overall security testing approach, providing different perspectives and uncovering unique vulnerabilities. By implementing these techniques together, organisations can achieve a more in-depth and robust security testing strategy that can help ensure their applications are secure from the multitude of threats present in today's cybersecurity landscape.

7.4.2. The Role of Penetration Testing in Application Security

Penetration testing, also known as pen testing, is a critical component in the process of securing applications. It involves simulating malicious attacks on systems, networks, or web applications to identify vulnerabilities that could be exploited by attackers. In application security, pen testing plays an essential role in uncovering weaknesses that may not be visible during static or dynamic analysis of an application's code or operation.

Penetration testing goes beyond automated tests and engages ethical hackers to actively probe and exploit vulnerabilities, thereby offering a realistic assessment of an application's security posture. This type of testing considers not just the underlying code but also the application's behaviours and its interactions with other systems, providing a more comprehensive evaluation of risk.

A crucial part of pen testing is its focused, intentional approach to identify vulnerabilities. Unlike other automated tests that scan for known issues, pen testing is more exploratory, probing the system for potential weak points that a motivated attacker might find and exploit. It also considers the potential business impact of the vulnerabilities discovered, classifying them based on their severity and the threat they pose to the organisation.

For instance, penetration testers might try to bypass security mechanisms, escalate privileges, or exploit vulnerabilities to gain unauthorised access. By simulating these attacks, organisations gain insights into potential attack vectors, allowing them to understand how an attacker might breach their application, the steps involved in such a breach, and the potential impact of a successful attack.

Pen testing can be performed manually or through automated tools, but typically involves a combination of both. Manual techniques allow for creative and unanticipated attack vectors, while automated tools ensure thoroughness and efficiency.

Another key advantage of penetration testing is its ability to test security controls and incident response mechanisms. It's one thing to have defences in place, but it's equally crucial to know how well these defences will hold up under a real-world attack. A pen test provides an opportunity to validate the effectiveness of these controls and evaluate the organisation's response to a detected incident.

While it might seem counterintuitive to invite an attack on your own applications, the value of penetration testing is precisely in its adversarial nature. By actively seeking out weaknesses and vulnerabilities, businesses can proactively address issues before they can be exploited by malicious actors, thereby reinforcing their security posture and strengthening trust with customers and stakeholders.

In conclusion, penetration testing forms a vital part of application security. With the continual evolution of cyber threats, the need for an in-depth, adversarial approach to testing has never been greater. By incorporating penetration testing into their security practices, organisations can stay one step ahead of attackers and ensure their applications are as secure as possible.

7.4.3. Continuous Security Monitoring and Auditing

Continuous security monitoring and auditing serve as pivotal processes in maintaining the integrity, confidentiality, and availability of applications in a dynamic digital environment. These processes offer proactive approaches to identifying, responding to, and preventing security threats, thereby fostering a more resilient application infrastructure.

Continuous security monitoring involves real-time collection, analysis, and alerting on various types of security-related information from applications and their environments. This goes beyond mere system health checks, diving into specifics like user behaviours, data access patterns, system vulnerabilities, and application performance issues. By doing so, it provides a comprehensive view of the application's security status, helping in timely identification and management of potential threats.

It is important to understand that continuous monitoring isn't only about technology. It incorporates a mix of automated tools and human analysis, bringing together various sources of data and transforming them into actionable intelligence. The insights gained from this monitoring can drive informed security decisions and aid in the quick mitigation of potential risks.

Continuous monitoring is also essential for maintaining regulatory compliance. Many industry standards and regulations require ongoing monitoring to ensure that controls are working as expected and that any deviations are quickly identified and remediated. Regular monitoring and timely reporting can aid organisations in demonstrating compliance during audits.

In conjunction with continuous monitoring, auditing is a critical security process that involves systematically reviewing and examining security procedures, controls, and activities. Security auditing assesses whether security policies are being adhered to, and whether they remain effective in the face of evolving threats and changing business needs.

Auditing can include various activities, such as reviewing user access logs, examining system configurations, evaluating incident response procedures, and assessing data protection measures. These audits can uncover non-compliance or operational inefficiencies and suggest areas for improvement.

However, the process of auditing should not be a one-off activity. With the constant evolution of cyber threats and changes in business processes, regular audits are necessary to ensure ongoing security. These audits need to be comprehensive, encompassing all aspects of application security – from coding practices and deployment processes to user access controls and data protection measures.

When combined, continuous monitoring and auditing form a robust security approach that not only detects threats but also assesses the effectiveness of security controls and procedures. It's a proactive stance, one that keeps a constant watch on the application's security posture, ensuring that organisations can stay ahead of the ever-evolving cyber threat landscape. It facilitates an environment where security becomes an integral part of the application lifecycle, and not just a reactive measure.

CHAPTER 8. CYBERSECURITY FRAMEWORKS AND STANDARDS

Cybersecurity frameworks and standards are essential tools in the arsenal of organisations aiming to ensure the security and integrity of their digital assets. They provide guidelines and best practices to help manage and reduce cybersecurity risks. This chapter, "Cybersecurity Frameworks and Standards," delves into the leading security frameworks and standards, their key components, how they can be implemented, and the roles they play in promoting an effective cybersecurity strategy.

Different organisations and sectors have varying needs when it comes to cybersecurity, due to which a one-size-fits-all approach is ineffective. Here, cybersecurity frameworks come into play, providing flexible strategies tailored to the unique needs of each entity. From small businesses to multinational corporations, hospitals to universities, government agencies to non-profit organisations, everyone can benefit from implementing a well-designed cybersecurity framework.

Many cybersecurity frameworks have been developed by renowned organisations, both governmental and non-governmental, including the National Institute of Standards and Technology (NIST), the International Organisation for Standardisation (ISO), and the Information Systems Audit and Control Association (ISACA). These frameworks are structured sets of guidelines that help an organisation manage and reduce cybersecurity risks by focusing on key areas such as risk management, access control, incident response, and ongoing monitoring.

Complementing these frameworks, cybersecurity standards offer specific rules and technical guidelines. They assist organisations in maintaining the security and integrity of their information systems. Standards often focus on specific areas such as data protection, network security, and incident management.

Understanding and implementing these frameworks and standards is critical in today's interconnected world, characterised by relentless cyber threats. They not only provide guidance on protecting an organisation's systems and data but also ensure compliance with legal and regulatory requirements related to cybersecurity.

In this chapter, we will explore some of the most prominent and widely used cybersecurity frameworks and standards, such as the NIST Cybersecurity Framework, ISO 27001, and the CIS Controls. We will delve into their structure, key components, and guidelines, and explore how they can be effectively implemented within an organisation. Furthermore, we will discuss the importance of these frameworks and standards in driving a culture of cybersecurity and how they can enhance an organisation's resilience against cyber threats.

The ultimate aim is to provide an understanding of how these frameworks and standards can be utilised as a roadmap to bolster cybersecurity posture and navigate the complex landscape of cyber threats and vulnerabilities. Whether your organisation is just starting on its cybersecurity journey or looking to enhance its existing defences, this chapter will offer invaluable insights into the world of cybersecurity frameworks and standards.

8.1. Overview of ISO 27001 and ISO 27002

The International Organisation for Standardisation (ISO) is recognised worldwide for its contribution in defining standards across a multitude of areas, including cybersecurity. ISO 27001 and ISO 27002, in particular, play a crucial role in helping organisations manage and improve their information security posture. In this section, we'll delve into the specifics of these two standards, elucidating their importance, structure, and the integral role they play within an information security management system (ISMS).

ISO 27001 is an international standard that provides a model for establishing, implementing, operating, monitoring, reviewing, maintaining, and improving an ISMS. The ISMS is an overarching management framework through which the organisation identifies, analyses, and addresses its information risks. The standard does not mandate specific information security controls but stops at the requirement that information security be addressed using a risk management process and giving assurance to stakeholders that risks are adequately managed.

Fundamentally, ISO 27001 underscores the significance of an organisation's commitment to information security at every level. The standard has a process-oriented approach, emphasising the implementation and maintenance of a robust ISMS. It also stresses the importance of continuous improvement, a crucial aspect of an effective ISMS, which ensures the system's longevity by adapting to changes in the business and threat landscape.

ISO 27002, on the other hand, complements ISO 27001 by providing a detailed, practical guide for implementing information security controls.

While ISO 27001 lays out the requirements for an ISMS, ISO 27002 provides a comprehensive set of controls that an organisation can adopt based on its unique risk landscape. These controls, grouped into various categories, encompass areas like information security policies, human resource security, access control, cryptography, physical and environmental security, operations security, and more.

ISO 27002 is often used in conjunction with ISO 27001 to help organisations identify and map the controls required for their ISMS. While ISO 27001 is certification-oriented, ISO 27002 provides actionable guidance for organisations, assisting them in selecting security controls based on the specific outcomes of their risk assessment and risk treatment processes.

The symbiotic relationship between ISO 27001 and ISO 27002 can't be overstated. Together, these standards provide a comprehensive and flexible approach towards managing information security risks. Organisations adopting these standards not only enhance their security posture but also inspire trust among stakeholders – a vital element in today's data-driven business world.

As we dive deeper into this section, we'll explore the essential elements of these standards, highlighting the key principles that underpin them, and offering guidance on how to implement and align them with your organisation's specific cybersecurity objectives. Regardless of your organisation's size or the sector it belongs to, understanding and applying these ISO standards is a definitive step towards bolstering your cybersecurity strategy and resilience.

8.1.1. Understanding ISO 27001: Information Security Management System

ISO 27001 is a globally recognised standard that provides a robust framework for implementing an Information Security Management System (ISMS) within an organisation. Issued by the International Organisation for Standardisation (ISO), it specifies the requirements for establishing, implementing, maintaining, and continuously improving an ISMS.

At its core, ISO 27001 mandates the systematic and structured approach towards managing information risks. It does so by encouraging the implementation of a suite of holistic and comprehensive security practices that are integrated into the daily operations of a business. The standard provides the methodology for risk assessment and risk treatment, a critical aspect of an ISMS, along with guidance on how to manage information security in a broader context.

ISO 27001 recognises that data is one of the most valuable assets a business possesses. As such, maintaining its confidentiality, ensuring its integrity, and guaranteeing its availability are of paramount importance. To this end, the ISMS serves as a systematic approach to managing sensitive company information so that it remains secure.

In terms of structure, ISO 27001 is segmented into main clauses and control categories, each catering to different aspects of information security management. The main clauses provide guidance on the context of the organisation, leadership, planning, support, operation, performance evaluation, and improvement. The control categories, on the other hand, provide a set of best practices on various areas, such as access control, operations security, information security incident management, and compliance.

269

ISO 27001 also emphasises a process of continuous improvement, adhering to the Plan-Do-Check-Act (PDCA) cycle. This iterative process ensures that the ISMS remains effective and suitable by continuously reviewing and improving the system, keeping pace with changes in the security threats, the business environment, and the organisational objectives.

The standard's risk-oriented approach requires the identification and assessment of security risks to the organisation's information. Risk assessments under ISO 27001 are performed systematically, repeatable, and consistently throughout the organisation, and relevant risks are assessed in terms of the potential impacts to the organisation and the likelihood of occurrence.

ISO 27001 requires the engagement of people across the organisation, from executive leadership who endorse the ISMS, to operations who implement the controls, to internal audit who verify compliance. This organisation-wide involvement fosters a culture of security and aids in the early identification and treatment of information security risks.

Despite its comprehensive nature, ISO 27001 is flexible and can be tailored to the needs of any organisation, irrespective of its size or the nature of its business. The standard provides a scalable and adaptable approach, allowing organisations to select controls as they see fit, based on their unique risk environment and business context.

When appropriately implemented, ISO 27001 provides numerous benefits to an organisation. Not only does it strengthen the organisation's information security posture, but it also demonstrates a commitment to a high level of security standards, thus fostering trust among customers, stakeholders, and regulators. Furthermore, having an ISO 27001 certification can be a strong differentiator in competitive markets, highlighting an organisation's dedication to safeguarding valuable data.

8.1.2. ISO 27002: Code of Practice for Information Security Controls

ISO 27002, also issued by the International Organisation for Standardisation (ISO), serves as a supplementary standard to ISO 27001 and provides a detailed set of best practices for establishing, implementing, and improving information security controls. It is a code of practice intended to be used as a reference by organisations and offers practical guidance on how to manage information security risks in the real world.

The objective of ISO 27002 is to give further detail to the control objectives and controls that are listed in the annex of ISO 27001. It takes the high-level requirements of ISO 27001 and supplements them with a comprehensive set of information security control objectives and controls (a total of 114 across 14 categories) that can be implemented based on the specific risks the organisation faces.

The 14 categories include areas such as security policies, organisation of information security, human resources security, asset management, access control, cryptography, physical and environmental security, operations security, communications security, system acquisition, development and maintenance, supplier relationships, information security incident management, information security aspects of business continuity management, and compliance.

For example, within the access control category, ISO 27002 provides guidance on how to manage user access through a formal user registration and de-registration process, how to manage privileged access rights, how to review user access rights, and how to manage secret authentication information, among others.

ISO 27002, however, does not provide a specification for an ISMS as ISO 27001 does, and organisations cannot be certified against ISO 27002. It is essentially a comprehensive catalogue of good practices, a toolbox from which organisations can select controls relevant to their specific risks as identified and assessed through an ISMS compliant with ISO 27001.

Another crucial aspect of ISO 27002 is its flexibility. While it provides a wealth of guidance on information security controls, it does not mandate that every control be implemented. The specific controls adopted and how they are employed are determined by several factors, including the organisation's specific business requirements, risk appetite, and the wider regulatory and legal environment.

Implementation of ISO 27002 can significantly enhance an organisation's overall information security posture. Its wide-ranging set of controls, when applied judiciously and in the context of a well-structured ISMS, helps protect an organisation's information assets by addressing potential vulnerabilities.

Moreover, it demonstrates to stakeholders, customers, and regulators that the organisation is committed to the rigorous practice of information security management. This commitment not only reduces the risk of security breaches but also can enhance an organisation's reputation, build customer trust, and result in a competitive advantage.

ISO 27002's role as a guidance document, therefore, is pivotal to the application of an ISMS under ISO 27001. The combination of these two standards offers a solid foundation for an organisation's information security management endeavours, with ISO 27001 providing the overarching framework and ISO 27002 offering the detailed, actionable practices.

8.1.3. Implementing ISO 27001 and ISO 27002 in Organisations

Implementing ISO 27001 and ISO 27002 within an organisation requires a methodical approach, encompassing a detailed understanding of the organisation's information security needs, a comprehensive risk assessment, and a structured implementation plan.

It starts with understanding the organisation's unique context. This means examining the organisation's objectives, information security requirements, legal and regulatory landscape, and the needs and expectations of interested parties. This step sets the scope of the ISMS (Information Security Management System), which is vital in determining the applicability of controls and risk management efforts.

Following this, a detailed risk assessment is undertaken as per ISO 27001 requirements. This involves identifying risks associated with the loss of confidentiality, integrity, and availability of information. These risks are then analysed and evaluated to decide on the risk treatment options. The results of this assessment drive the selection of appropriate controls from ISO 27002 to mitigate these risks.

Next, the control objectives and controls from ISO 27002 are implemented in the context of the organisation's ISMS. This step typically involves creating and documenting policies, procedures, and processes aligned with the guidance of ISO 27002. Here, a risk treatment plan is developed which outlines the necessary actions, resources, responsibilities, and priorities for managing information security risks.

A key point to note here is that not all the controls outlined in ISO 27002 need to be applied. Instead, the standard recommends that organisations select, and tailor controls based on their risk assessment and business requirements. Essentially, ISO 27002 acts as a toolbox of potential measures, from which the most suitable ones are chosen, based on the specific needs of the organisation.

It is crucial to provide training and awareness programs for staff during this phase, ensuring they understand their roles and responsibilities, as well as the relevance of information security to their everyday tasks. This human-centric approach can help to foster a culture of security within the organisation.

Once the ISMS is in place, the organisation must monitor and review its performance regularly. This includes conducting internal audits and management reviews to ensure the ISMS remains effective and continually improves. Here, the focus is on checking if the controls are working as intended, identifying areas for improvement, and making necessary adjustments.

Moreover, an important part of implementing ISO 27001 is preparing for certification by an accredited body if the organisation chooses to do so. While ISO 27002 does not have a certification scheme, the successful implementation of its control set contributes to achieving ISO 27001 certification, demonstrating that the organisation's ISMS meets international standards.

In conclusion, implementing ISO 27001 and ISO 27002 is a strategic decision that requires commitment, resources, and ongoing effort. It's not a one-time project but rather an ongoing process that becomes part of the organisation's culture and operations, contributing to resilience against the evolving landscape of information security threats.

8.2. Understanding the NIST Cybersecurity Framework

The National Institute of Standards and Technology (NIST) Cybersecurity Framework is a voluntary guidance, based on existing standards, guidelines, and practices, designed to help organisations better manage and reduce cybersecurity risk. Developed through a collaborative process involving industry, academia, and government agencies, the framework was released in 2014 in response to an executive order issued by the President of the United States to enhance the security of critical infrastructure.

The NIST Cybersecurity Framework is unique in its focus on risk management and its applicability across various industries and organisations of all sizes. It provides a common language for understanding, managing, and expressing cybersecurity risk, both internally and externally. Its flexibility allows it to be applied to a wide range of sectors and organisations, from small businesses to large enterprises, and across many industries.

The framework is designed around three primary components: the Framework Core, the Framework Profile, and the Framework Implementation Tiers.

The Framework Core is a set of cybersecurity activities, desired outcomes, and applicable references that are common across critical infrastructure sectors. It provides detailed guidance for developing individual organisational Profiles. Through the use of the Core, an organisation can align its cybersecurity activities with its business requirements, risk tolerances, and resources.

The Framework Profile is essentially the alignment of the Functions, Categories, and Subcategories with the business requirements, risk tolerance, and resources of the organisation. A Profile can be used to describe both the current state and the desired state of specific cybersecurity activities, which can reveal gaps that should be addressed to meet cybersecurity risk management objectives.

The Framework Implementation Tiers assist organisations in viewing and expressing their management of cybersecurity risk and the processes in place to manage that risk. Tiers range from Partial (Tier 1) to Adaptive (Tier 4) and describe an increasing degree of rigor and sophistication in cybersecurity risk management practices and the extent to which cybersecurity risk management is informed by business needs.

The NIST Cybersecurity Framework has been widely adopted across various sectors due to its flexibility and comprehensiveness. By offering a prioritised, flexible, repeatable, performance-based, and cost-effective approach, it helps organisations manage their cybersecurity-related risk in a climate of evolving threats, providing a high level of security to protect business operations and ultimately contribute to their bottom line. It provides a road map for improving an organisation's cybersecurity posture, making the digital infrastructure more robust, resilient, and secure.

8.2.1. The Five Functions of NIST Framework

The NIST Cybersecurity Framework, a robust guide to managing and mitigating cybersecurity risk, revolves around five crucial functions: Identify, Protect, Detect, Respond, and Recover. These functions are not intended to form a sequential path, but rather to offer a strategic view of the lifecycle of an organisation's management of cybersecurity risk.

1. Identify: The initial step in the NIST Cybersecurity Framework, identification, involves comprehending the business context, the assets at risk, and the associated risks. This could range from understanding the business environment, governance, risk assessment, risk management strategy, and supply chain risks. The Identify function serves to develop an organisational understanding to manage cybersecurity risk to systems, people, assets, data, and capabilities. Effective identification aids in making informed decisions and prioritising efforts in line with the business's risk management strategy and its risk tolerance threshold.

2. Protect: This function focuses on the safeguarding measures that are put in place to protect the critical infrastructure services of the organisation. This includes access control, awareness and training, data security, information protection processes and procedures, maintenance, and protective technology. The aim is to develop and implement the appropriate safeguards to ensure delivery of critical infrastructure services, thereby reducing the impact of potential cybersecurity events.

3. Detect: This function is concerned with the implementation of appropriate activities to identify a cybersecurity event promptly. Organisations must monitor their systems and assets to detect potential cybersecurity events continually. Anomalies and events are identified; cybersecurity events are detected; and their impact is assessed. The early detection of a cybersecurity incident can significantly reduce the potential harm caused by the event.

4. Respond: This function focuses on the actions that are taken in response to a detected cybersecurity event. Activities under this function include response planning, communications, analysis, mitigation, and improvements. The objective is to develop and implement the appropriate activities to take action regarding a detected cybersecurity event. It ensures the ability to contain the impact of a potential cybersecurity event.

5. Recover: This final function addresses the activities necessary to restore any capabilities or services that were impaired due to a cybersecurity event. This includes recovery planning, improvements, and coordination with external parties. The aim is

to develop and implement appropriate activities to maintain plans for resilience and to restore any capabilities or services that were impaired due to a cybersecurity event.

Each of these five functions is critical to an effective cybersecurity strategy, and each includes a variety of categories and subcategories that specify outcomes aligned with industry standards and best practices. Organisations can leverage these functions, depending on their specific needs, risks, and objectives. It's vital to note that the NIST Cybersecurity Framework's five functions do not entail a rigid sequence to be followed. Instead, they present a dynamic process where all functions must be concurrently and continuously addressed. This approach fosters an active understanding of cybersecurity risk and helps maintain a resilient posture that can rapidly adapt and respond to changes in the business and technological environment.

8.2.2. Applying the NIST Framework: A Case Study

Applying the NIST Cybersecurity Framework requires a clear understanding of the organisation's risk tolerance, resource availability, and security requirements. The success of such application can be illustrated through a case study of a medium-sized healthcare provider, "HealthSecure," which faced a series of cybersecurity challenges.

HealthSecure, like most healthcare providers, had vast amounts of sensitive data. From patient records to financial information, the volume and value of this data made them a prime target for cyber threats. Despite implementing basic security measures, HealthSecure experienced several minor cyber incidents, raising concerns about the potential for a significant breach. They decided to implement the NIST Cybersecurity Framework to enhance their cybersecurity posture.

Step 1: Identify: HealthSecure started by assessing its business context, identifying the critical services they provided, the data necessary for these services, and the technologies used. They catalogued their digital assets, including hardware, data, systems, and networks, and identified associated risks. They understood their risk tolerance and prioritised their assets based on their importance to service delivery.

Step 2: Protect: With a clear understanding of their assets and risks, HealthSecure developed safeguards. They implemented stricter access controls, ensuring only authorised personnel could access sensitive data. They developed a cybersecurity awareness and training program for all staff. Additionally, they established data security policies and protective technology, including regular system backups, data encryption, and antivirus software.

Step 3: Detect: To promptly identify cybersecurity events, HealthSecure established continuous monitoring. They set up a Security Operations Centre (SOC) to monitor system logs and network traffic for abnormal activities. They also established an incident detection and assessment process to identify potential cybersecurity events.

Step 4: Respond: HealthSecure developed a response plan detailing the steps to take upon detecting a cybersecurity event. The plan included communication procedures, including notification of affected parties and regulatory bodies. It also involved mitigation strategies to limit the impact of the event and strategies for preserving evidence for post-event analysis.

Step 5: Recover: Finally, HealthSecure established a recovery plan to restore any capabilities or services impaired due to a cybersecurity event. This plan included data and system restoration procedures from backups, improvements to avoid recurrence of similar incidents, and coordination with external parties, including cyber insurance providers and law enforcement.

Over the next year, HealthSecure noticed a reduction in cybersecurity incidents. When incidents did occur, they were detected and mitigated faster, reducing the overall impact. By applying the NIST Cybersecurity Framework, HealthSecure improved its risk management and bolstered its resilience to cybersecurity threats.

This case study demonstrates the practical application of the NIST Cybersecurity Framework and underscores the value it provides in improving an organisation's cybersecurity posture. However, it's important to remember that cybersecurity is a continuous process. Regular reviews and updates to the organisation's cybersecurity approach are crucial to ensure continued efficacy in the face of evolving cyber threats.

8.3. PCI DSS for Cardholder Data Security

In an increasingly digital world, securing cardholder data is of paramount importance for businesses of all sizes. This data, often used for financial transactions, is a prime target for cybercriminals who aim to exploit vulnerabilities in payment card processing systems. Failure to adequately protect this sensitive data can lead to significant financial and reputational damages. To address this, the Payment Card Industry Data Security Standard (PCI DSS) was established as a comprehensive set of requirements designed to ensure that all companies that process, store, or transmit cardholder data maintain a secure environment.

Created by the Payment Card Industry Security Standards Council (PCI SSC), a consortium of major credit card companies, PCI DSS provides a baseline of technical and operational requirements designed to protect cardholder data. The Standard applies to all entities involved in payment card processing—including merchants, processors, acquirers, issuers, and service providers—as well as all other entities that store, process, or transmit cardholder data. The PCI DSS includes requirements for security management, policies, procedures, network architecture, software design, and other critical protective measures.

PCI DSS is not just about compliance; it's about protecting businesses and their customers. By adhering to these requirements, companies can protect their reputation, reduce the risk of data breaches, and potentially avoid costly fines and penalties associated with non-compliance. The nature and volume of cardholder data can vary considerably between organisations, making the implementation of PCI DSS a unique journey for each. However, regardless of size, the importance of adhering to PCI DSS cannot be overstated.

With twelve overarching requirements grouped into six control objectives, the PCI DSS provides a multifaceted security standard that covers every aspect of the cardholder data environment. From constructing and maintaining a secure network, to implementing strong access control measures, regular monitoring, and testing of networks, and maintaining an information security policy, the guidelines provide a solid foundation for security.

The PCI DSS is updated periodically to address emerging threats and changes in the market. It underscores the importance of ongoing security rather than one-time compliance, prompting organisations to embed data security into their everyday operations.

In the subsequent sections of this chapter, we delve into the intricacies of the PCI DSS, including an understanding of the requirements, how to implement them, and the benefits of compliance. This information is crucial to any organisation aiming to secure cardholder data effectively. Whether you're a small online retailer or a large multinational corporation, understanding and implementing the PCI DSS is a critical step towards robust data security.

8.3.1. Understanding PCI DSS Requirements

The Payment Card Industry Data Security Standard (PCI DSS) comprises twelve essential requirements designed to secure cardholder data. These requirements, developed by the founding payment brands of the PCI Security Standards Council, include American Express, Discover, JCB, MasterCard, and Visa, apply to any organisation that stores, processes, or transmits cardholder data. Here's a comprehensive view of what these requirements entail.

Requirements 1 and 2 form the bedrock of PCI DSS, revolving around the construction and maintenance of a secure network. Requirement 1 mandates the installation and maintenance of a firewall configuration to protect cardholder data. Firewalls are a primary line of defence in any network security infrastructure, inspecting incoming and outgoing network traffic based on predefined security rules, blocking traffic that does not comply with these rules. As such, they are crucial in preventing unauthorised access to network resources.

Requirement 2 is about not using vendor-supplied defaults for system passwords and other security parameters. Default usernames and passwords for technology products are widely known and freely available on the internet, providing an easy access route for cybercriminals. Changing these is fundamental to protecting your systems from unauthorised access.

The next two requirements, 3 and 4, focus on protecting stored cardholder data and ensuring the encryption of transmission of this data across open, public networks. Requirement 3 pertains to maintaining a data storage policy, including not storing sensitive data unnecessarily and encrypting it where it is stored. Requirement 4 covers the transmission of cardholder data across open, public networks, requiring it to be encrypted to prevent unauthorised interception during transmission.

Requirements 5 and 6 mandate the use of antivirus software and the development of secure systems and applications. Requirement 5 requires the use of antivirus software on all systems commonly affected by malware to protect systems from known threats. Requirement 6 revolves around the secure development lifecycle, including a process for identifying security vulnerabilities and patching them in a timely manner.

Requirements 7, 8, and 9 focus on implementing strong access control measures. These include restricting access to cardholder data by business need-to-know (Requirement 7), identifying and authenticating access to system components (Requirement 8) and restricting physical access to cardholder data (Requirement 9). These measures ensure that only authorised individuals can access cardholder data, whether electronically or physically.

The last set of requirements, 10, 11, and 12, pertain to regular monitoring and testing of networks and maintaining a robust information security policy. Requirement 10 requires the tracking and monitoring of all access to network resources and cardholder data to detect and respond to unauthorised access promptly. Requirement 11 involves regularly testing security systems and processes to identify vulnerabilities that could potentially be exploited. Lastly, Requirement 12 necessitates a policy that addresses information security for all personnel to ensure a clear understanding of the corporate posture and the importance of security in their roles.

In summary, PCI DSS requirements establish a comprehensive framework that prioritises security across all aspects of cardholder data handling. By understanding these requirements, organisations can more effectively protect themselves and their customers from data breaches and maintain the trust necessary for successful business operations. Implementing these standards isn't just about compliance; it's a robust strategy for risk management and overall business health.

8.3.2. Achieving and Maintaining PCI DSS Compliance

Achieving and maintaining PCI DSS (Payment Card Industry Data Security Standard) compliance is a critical task for any organisation that handles cardholder data. It's an ongoing process that involves continuous monitoring, reporting, and improving of security measures to protect against evolving cybersecurity threats. Below is an overview of the process for achieving and maintaining compliance with the PCI DSS.

Firstly, it's important to understand that achieving PCI DSS compliance starts with a comprehensive risk assessment. This process involves identifying all the systems, processes, and personnel that interact with or could potentially impact the security of cardholder data. This includes but is not limited to payment systems, databases, networks, endpoints, third-party service providers, and employees.

Following the risk assessment, the organisation should map out the data flows of cardholder data across its systems and processes. This exercise helps in identifying potential vulnerabilities and weaknesses that could be exploited by attackers. It also aids in understanding the extent of applicability of PCI DSS requirements, as these requirements apply to all system components included in or connected to the cardholder data environment.

Once the organisation understands its cardholder data environment, the next step is to implement the necessary controls to meet the twelve core requirements of the PCI DSS. This stage could require considerable effort and resources, as it may entail implementing new technologies, reconfiguring existing systems, developing new processes, and training personnel. Given the wide-ranging nature of the PCI DSS requirements, organisations often prioritise their efforts based on the identified risks and their potential impact.

With the controls in place, the organisation should conduct testing to verify that the controls are effective, and the organisation meets the PCI DSS requirements. This stage often involves engaging a Qualified Security Assessor (QSA) or using an internal security assessor (ISA), depending on the organisation's transaction volume and the payment brands' requirements.

Upon successful validation of compliance, the organisation receives a Report on Compliance (ROC) or an Attestation of Compliance (AOC), which can be provided to acquiring banks and payment brands as proof of compliance. However, it's critical to understand that PCI DSS compliance is not a one-time event. It's a continuous process of monitoring, reviewing, and improving the security controls and processes.

Maintaining PCI DSS compliance involves regular monitoring and logging of systems and networks to detect anomalies or security incidents that could indicate a compromise of cardholder data. Organisations should regularly review their logs and security alerts to identify and respond to potential security incidents promptly. They should also conduct regular vulnerability scans and penetration tests to identify and remediate vulnerabilities.

Another crucial aspect of maintaining PCI DSS compliance is staying abreast of changes in the standard and the wider payment security landscape. The PCI Security Standards Council periodically updates the PCI DSS to address emerging threats and challenges, so organisations should ensure they understand and implement any new requirements when they are released.

Lastly, maintaining PCI DSS compliance requires an ongoing commitment to security awareness and training among employees. As humans are often the weakest link in security, it's important that all personnel understand their role in maintaining PCI DSS compliance and the potential consequences of non-compliance.

In conclusion, achieving and maintaining PCI DSS compliance requires a structured and disciplined approach towards risk assessment, implementation of controls, regular testing, and continuous monitoring. By meeting these requirements, organisations can significantly reduce the risk of a data breach, protect their reputation, and earn the trust of their customers.

8.4. Health Insurance Portability and Accountability Act (HIPAA) for Healthcare Data

Healthcare is one of the most important sectors that deal with sensitive and personal information. Among the data that it handles are health records, which are rich in personal, sensitive, and highly valuable information. As such, protecting this data from breaches and unauthorised access has become a top priority for the industry. One of the key regulations guiding healthcare data protection in the United States is the Health Insurance Portability and Accountability Act (HIPAA).

HIPAA was enacted in 1996, originally aiming to make it easier for people to keep health insurance and protect the confidentiality and security of healthcare information. Over time, with the digital revolution, HIPAA's scope has expanded to cover the security and privacy aspects of electronic health records (EHRs), shaping the way healthcare providers manage and secure patient data.

Under the umbrella of HIPAA, there are two critical rules to understand: the Privacy Rule and the Security Rule. The Privacy Rule, as its name suggests, is focused on the right of an individual's health information privacy, specifying what constitutes protected health information (PHI), and stipulating how covered entities must handle it. The Security Rule complements the Privacy Rule by providing a detailed outline of the three types of safeguards - administrative, physical, and technical - that organisations must have in place to secure individuals' electronic protected health information (e-PHI).

For entities under its purview, such as health plans, healthcare clearinghouses, and healthcare providers that transmit any health information electronically, HIPAA is not optional. Non-compliance can lead to hefty penalties. Moreover, the reputation damage from HIPAA violations can have lasting effects on a healthcare organisation's credibility and patient trust.

HIPAA's goal is to strike a balance between sharing information to improve patient care and safeguarding individual privacy rights. While HIPAA compliance can be challenging due to its broad scope and complex requirements, it is crucial in maintaining the privacy and security of sensitive healthcare data. The subsequent sections will delve deeper into understanding HIPAA's key provisions, implementation strategies, and best practices for maintaining compliance.

In the era of increasing digitisation, especially in healthcare where technologies such as telemedicine, artificial intelligence, and machine learning are making significant inroads, understanding and complying with regulations like HIPAA has never been more critical. Let's embark on this journey to demystify HIPAA and the fundamental role it plays in healthcare data security.

8.4.1. Understanding HIPAA Privacy and Security Rules

The Health Insurance Portability and Accountability Act (HIPAA) revolves around two primary regulations: the Privacy Rule and the Security Rule. Both of these rules, with distinct but complementary objectives, form the cornerstone of how healthcare data is protected in the United States.

The Privacy Rule came into effect first, establishing national standards for the protection of health information. At its core, the Privacy Rule addresses the use and disclosure of individuals' health information, known as

Protected Health Information (PHI). PHI refers to any information held by a covered entity that concerns health status, provision of health care, or payment for healthcare and can be linked to an individual. This includes any part of a patient's medical record or payment history. The Privacy Rule stipulates that such PHI should be appropriately protected while allowing for necessary health information flow to promote high-quality healthcare and protect public health.

The Privacy Rule applies to organisations classified as 'covered entities'. These entities include healthcare providers, health plans, and healthcare clearinghouses. Additionally, business associates, or entities that offer services to covered entities that involve access to PHI, are also obliged to follow HIPAA rules. The Privacy Rule mandates that covered entities must put safeguards in place to protect patient information and ensure PHI is disclosed only to the patient or other entities with the patient's explicit permission.

The Security Rule, on the other hand, was created to specifically address Electronic Protected Health Information (e-PHI). It outlines three types of security safeguards required for compliance: administrative, physical, and technical. Administrative safeguards involve policies and procedures designed to show how the entity will comply with HIPAA, while physical safeguards focus on physical access to PHI. Technical safeguards involve the technology that protects PHI and controls access to it.

For instance, technical safeguards include practices like using automatic logoff to terminate sessions after a period of inactivity or encryption of e-PHI to prevent unauthorised access. Physical safeguards may include policies on workstation use or the disposal and reuse of electronic media, while administrative safeguards could involve risk analysis and management policies or workforce training and management.

The Security Rule is flexible and allows for scalability, meaning that the security measures implemented by an entity can vary based on factors such as the entity's size, complexity, or capabilities. However, what remains non-negotiable is that all covered entities must analyse their risks and establish reasonable and appropriate security measures.

HIPAA's Privacy and Security Rules work hand in hand to ensure the confidentiality, integrity, and availability of all e-PHI an entity creates, receives, maintains, or transmits. While the Privacy Rule broadly covers all PHI, including paper and electronic, the Security Rule zooms in on safeguarding e-PHI.

The increased digitisation of healthcare services, compounded by a rise in cybersecurity threats, makes these rules highly crucial in today's landscape. The nuanced understanding of HIPAA Privacy and Security Rules not only aids healthcare entities to stay compliant but also significantly helps in building trust and ensuring the safety of individuals' health information in an increasingly interconnected world. The following sections will provide more details about how to achieve compliance with these regulations effectively.

8.4.2. Achieving HIPAA Compliance: Steps and Strategies

Achieving compliance with the Health Insurance Portability and Accountability Act (HIPAA) can be a complex task, but it is a crucial one for organisations handling protected health information (PHI). Compliance not only mitigates the risk of data breaches and penalties but also builds trust with customers and partners. The following steps and strategies can guide organisations towards successful HIPAA compliance.

1. Understand the HIPAA Rules:

The first step towards HIPAA compliance is understanding the regulations themselves. This includes the Privacy Rule, the Security Rule, the Breach Notification Rule, and the Omnibus Rule. These regulations outline the rights of patients and the responsibilities of covered entities and their business associates. Understanding these rules will form the foundation of any compliance strategy.

2. Conduct a Risk Analysis:

HIPAA requires covered entities to conduct a risk analysis to identify potential vulnerabilities to the confidentiality, integrity, and availability of e-PHI. This involves evaluating data storage methods, IT infrastructure, and potential threats, both internal and external. The risk analysis should be thorough and should cover every area where PHI is stored, accessed, or transmitted.

3. Implement Administrative, Physical, and Technical Safeguards:

HIPAA's Security Rule mandates that covered entities implement appropriate safeguards in three areas: administrative, physical, and technical. Administrative safeguards involve policies and procedures to manage the selection, development, implementation, and maintenance of security measures. Physical safeguards protect physical computer systems and related buildings and equipment from hazards and unauthorised intrusion. Technical safeguards involve the technology that protects PHI and controls access to it. These might include encryption, access controls, audit controls, and integrity controls.

4. Develop Policies and Procedures:

Covered entities should develop and implement policies and procedures to comply with the HIPAA regulations. This should include a privacy policy, a security policy, a breach notification policy, and a policy for handling patient rights requests. Staff should be trained on these policies and procedures to ensure they understand their responsibilities.

5. Train Your Staff:

Regular training is vital to ensure that all staff members understand the HIPAA requirements and the organisation's specific policies and procedures. Training should cover a broad range of topics, including how to handle PHI, how to respond to a potential breach, and how to safeguard e-PHI. It's also essential to maintain documentation of all training activities.

6. Regular Auditing and Monitoring:

Regular auditing and monitoring can help organisations identify and address any non-compliance issues proactively. Audits should assess compliance with all aspects of HIPAA, including the Privacy Rule, the Security Rule, and the Breach Notification Rule. Any identified issues should be addressed promptly to mitigate potential risks.

7. Establish a Plan for Responding to Breaches:

Despite best efforts, breaches may still occur. Organisations should have a robust response plan in place, including identifying and assessing the breach, containing and rectifying the issue, notifying the necessary parties, and implementing measures to prevent future breaches.

8. Regular Review and Updates:

Lastly, achieving HIPAA compliance is not a one-time task. Regular reviews and updates of policies, procedures, risk analysis, and staff training are required to maintain compliance and to keep pace with changes in technology, threats, and business processes.

In conclusion, achieving HIPAA compliance requires a structured and continuous approach. While the process can be challenging, the benefits of protecting sensitive patient data and avoiding potential penalties and reputational damage far outweigh the investment of time and resources.

8.5. General Data Protection Regulation (GDPR) for EU Residents' Data

Navigating the complex landscape of cybersecurity regulations is a critical part of modern business, and it has become even more paramount with the advent of global data protection regulations such as the European Union's General Data Protection Regulation (GDPR). The GDPR, implemented in May 2018, revolutionised data protection standards and has shaped the digital landscape, affecting not only EU organisations but also any entity worldwide that processes the personal data of EU residents.

The GDPR has elevated data protection to a fundamental human right and imposes stringent rules on data handling and security. It mandates that organisations adhere to principles of data minimisation, purpose limitation, and confidentiality. Importantly, it has brought into sharp focus the concept of 'privacy by design and default', which ensures that privacy is an integral part of system design, rather than an afterthought.

This chapter will explore GDPR as a comprehensive legal framework aimed at the protection of EU residents' data. It introduces the concept of GDPR, its implications for organisations worldwide, and the significance of its role in data protection in an age where data breaches and privacy concerns are on the rise.

A key emphasis of the GDPR is on transparency and accountability. It requires organisations to be transparent about their data processing activities and to demonstrate their compliance. To this end, the GDPR introduces significant changes in data management practices, requiring organisations to incorporate data protection into their business models and operations at a fundamental level.

The GDPR's scope extends far beyond the borders of the EU. The regulation applies to any organisation, anywhere in the world, which processes the personal data of EU residents. This broad applicability means that global businesses must be aware of and comply with the GDPR or face potentially severe financial penalties.

GDPR compliance is not just about avoiding penalties, however. It also represents a significant opportunity for businesses. By demonstrating robust data protection practices, organisations can enhance their reputation, build trust with customers, and gain a competitive edge. Furthermore, the principles and practices underpinning GDPR compliance – such as data minimisation, purpose limitation, and data accuracy – can also help organisations streamline their data management processes, leading to efficiencies and improved decision-making.

In the subsequent sections of this chapter, we will delve deeper into the specific aspects of GDPR, including its key principles, rights of data subjects, and obligations of data controllers and processors. We will also explore the practical steps that organisations can take towards achieving GDPR compliance and the technologies that can support this journey. Furthermore, we will highlight some of the challenges and potential future developments in the data protection landscape under the GDPR.

As we traverse the intricate contours of GDPR, we should keep in mind that the GDPR is more than just a regulatory requirement. It is a significant step towards a world where the right to privacy and data protection is universally recognised and upheld. Understanding and complying with the GDPR is not just about doing business in the EU; it is about doing business ethically and responsibly in the digital age.

8.5.1. Key Provisions and Rights Under the GDPR

The General Data Protection Regulation (GDPR) introduced an unprecedented set of rights and provisions, altering the landscape of data privacy and security worldwide. This landmark legislation heralded a new era of personal data governance, focusing on privacy as a human right and empowering individuals with greater control over their data.

At the heart of the GDPR is the principle of personal data protection. Personal data is defined broadly, covering any information relating to an identified or identifiable natural person (known as the "data subject"). This could be anything from a name, a photo, an email address, bank details, posts on social networking websites, medical information, or even a computer's IP address.

A key provision of the GDPR is the requirement for clear consent. Under the GDPR, organisations must obtain explicit consent from data subjects before collecting, processing, or storing their data. This consent must be freely given, specific, informed, and unambiguous. It represents a significant shift from previous practices, where organisations could often rely on implied or passive consent. The onus is now on organisations to demonstrate that valid consent was obtained.

The GDPR has also introduced a suite of rights for data subjects, often referred to as the eight rights of the individual. These rights include:

1. The Right to Be Informed: This relates to the transparency of data processing activities. Organisations must provide clear and accessible information about how personal data is used.

2. The Right of Access: Individuals have the right to access their personal data and supplementary information. They have the right to know what data is being processed and how it is being processed.

3. The Right to Rectification: If personal data is inaccurate or incomplete, individuals have the right to have it corrected.

4. The Right to Erasure (Right to Be Forgotten): In certain circumstances, individuals can request the deletion or removal of personal data.

5. The Right to Restrict Processing: Individuals can request that the processing of personal data be restricted.

6. The Right to Data Portability: This allows individuals to obtain and reuse their personal data across different services.

7. The Right to Object: In certain circumstances, individuals have the right to object to the processing of their personal data.

8. Rights in relation to automated decision making and profiling: The GDPR has provisions to protect individuals against decisions made without human intervention.

The GDPR's emphasis on these rights signifies a shift in the power dynamic between organisations and individuals, giving the latter more control over their data. This shift requires organisations to rethink their data handling practices and prioritise the individual's rights in their operations.

Notably, the GDPR introduces significant penalties for non-compliance. Fines can reach up to €20 million or 4% of the company's global annual turnover, whichever is higher. The severity of these penalties underscores the importance of the legislation and serves as a strong incentive for organisations to take their data protection responsibilities seriously.

In conclusion, the key provisions and rights under the GDPR represent a comprehensive and robust framework for data protection. They encourage responsible data handling practices, foster trust between organisations and individuals, and contribute to the establishment of a privacy-centric digital culture. Despite the challenges of compliance, understanding and adhering to these provisions offers organisations an opportunity to enhance their reputation, build customer trust, and ultimately, ensure the long-term sustainability of their operations in an increasingly data-driven world.

8.5.2. Achieving GDPR Compliance: A Practical Guide

Achieving GDPR compliance is not a one-time exercise but a continuous endeavour that requires commitment and diligence. Here, we'll discuss a practical guide that provides steps towards achieving and maintaining GDPR compliance.

1. **Understand Your Data:** Begin by conducting a data audit to identify what personal data you collect, where it's stored, how it's used, and with whom it's shared. This includes understanding the lifecycle of the data: from collection to processing, storage, and eventual deletion. Such an understanding forms the foundation of your compliance efforts.

2. **Implement Data Protection Principles:** Embed GDPR's core principles in your data handling practices. These principles include lawfulness, fairness, transparency, purpose limitation, data minimisation, accuracy, storage limitation, integrity, and confidentiality.

3. **Obtain Clear Consent:** The GDPR insists on explicit and informed consent from data subjects. This means your terms and conditions must be clearly understandable and not hidden in legalese. Consent mechanisms should be user-friendly, providing individuals the freedom to opt-in and opt-out easily.

4. **Respect Data Subject Rights:** Create mechanisms that respect and facilitate the eight rights of individuals outlined in the GDPR. This includes mechanisms for data access, rectification, erasure, restriction of processing, data portability, and objection to processing.

5. **Ensure Data Security:** Implement appropriate technical and organisational measures to ensure the security of personal data. This could involve encryption, pseudonymisation, access control, secure data transfers, regular security audits, and having procedures in place to detect, report, and investigate data breaches.

6. **Appoint a Data Protection Officer (DPO):** Larger organisations or those processing sensitive data on a large scale may need to appoint a DPO. The DPO's role is to oversee data protection strategy and GDPR compliance.

7. **Privacy Impact Assessment (PIA):** If your processing activities are likely to result in high risk to individuals' rights and freedoms, conduct a PIA. This is a process that helps identify and mitigate these risks.

8. **Establish Data Breach Procedures:** GDPR requires organisations to report certain types of data breaches to the relevant supervisory authority within 72 hours of becoming aware of it. Ensure you have robust breach detection, investigation, and internal reporting procedures in place.

9. **International Data Transfers:** If you transfer personal data outside the European Economic Area (EEA), ensure you do so lawfully and protect the data appropriately.

10. **Staff Training and Awareness:** GDPR compliance is a collective responsibility. Regular staff training helps embed compliance into the everyday workings of your organisation and helps prevent data breaches resulting from human error.

Achieving GDPR compliance can be a significant task, particularly for larger organisations or those handling large quantities of sensitive data. But with a structured approach, it becomes manageable. Not only does GDPR compliance protect you from heavy fines and reputational damage, but it also signals to customers and stakeholders that you take data protection seriously. This trust and transparency can ultimately lead to a competitive advantage in the modern, data-driven marketplace.

CHAPTER 9. IMPLEMENTING A CYBERSECURITY PROGRAM

After traversing the multifaceted domains of cybersecurity, it is now time to utilise this knowledge to build a comprehensive cybersecurity program for your organisation. The goal is to fortify your organisation's digital infrastructure, protect sensitive data, and ensure business continuity in the face of potential cyber threats.

In a hyper-connected digital world, organisations operate within a vast, dynamic, and complex cybersecurity ecosystem. Each node, link, and pathway in this networked system can introduce potential vulnerabilities. These vulnerabilities, if not adequately addressed, can be exploited, leading to security breaches that could potentially cripple an organisation's operations and erode trust among stakeholders. To prevent this, a robust cybersecurity program must be put in place, providing the organisation with a formidable shield against a myriad of cyber threats.

This chapter will take you through the methodical process of establishing a cybersecurity program. This includes laying the groundwork, defining the structure, and putting into action the plans, policies, procedures, and solutions that make up the program. This journey, though demanding, is of critical importance, as it is about more than just defending against cyber threats. It's about instilling a culture of cybersecurity across the organisation, fostering digital trust and resilience.

First, we will outline the initial steps of planning and preparing for the program. This includes conducting a risk assessment, setting objectives, and defining the scope of the program. It also includes considerations for regulatory compliance and alignment with industry standards.

Next, we explore how to structure the program and define roles and responsibilities. A well-defined structure and clear accountability are crucial to the success of any cybersecurity program. This section will also discuss the significance of leadership commitment in fostering a security-conscious culture.

Following the organisational aspect, we delve into the strategic components of the program, including policy development, security awareness training, and implementing defence-in-depth strategies. This section underscores the need for comprehensive security policies, regular employee training, and multi-layered defence mechanisms.

Finally, we discuss how to maintain, monitor, and continuously improve the cybersecurity program. This part emphasises the iterative nature of cybersecurity. We discuss how continuous monitoring, regular audits, and timely updates help keep the program effective and relevant in an ever-evolving threat landscape.

Implementing a cybersecurity program is a significant but worthwhile investment. A well-designed and effectively managed cybersecurity program not only protects against immediate threats but also prepares the organisation for the future. As the digital world continues to evolve, and as new threats emerge, organisations will find that their cybersecurity program is not just a protective shield, but a strategic asset that enables them to navigate the digital future with confidence.

9.1. Conducting a Risk Assessment

Section 9.1 of this chapter delves into a crucial initial step in the journey of implementing a cybersecurity program - conducting a Risk Assessment. This process is integral as it provides organisations with a clear understanding of their existing vulnerabilities and the potential threats they may face. The aim is to identify, analyse, and evaluate the risks that could impact an organisation's information assets and operations.

In an era where technology is deeply embedded into the fabric of business operations, a myriad of risks arises from different fronts. These risks could emerge from hardware failures, software vulnerabilities, human errors, malicious attacks, or even natural disasters. Each of these can significantly disrupt an organisation's functioning, lead to data loss, and damage its reputation. Therefore, to strategise a comprehensive cybersecurity program, organisations must first determine what they are up against. This is the essence of conducting a risk assessment.

Risk assessments can be a complex process, but it is also a rewarding one. It is like embarking on an exploratory journey, delving into the organisation's digital ecosystem to uncover potential pitfalls. This journey encompasses numerous components, from identifying the organisation's valuable assets and recognising potential threats, to understanding vulnerabilities and estimating the impact of possible breaches.

During this process, the organisation must scrutinise its information systems and their environments, including software applications, databases, servers, networks, and endpoints. This analysis must also factor in the human element, considering potential lapses like weak passwords, phishing susceptibility, and lack of awareness about safe online practices.

The risk assessment also includes analysing the external risk environment. This means understanding the landscape of potential threats, such as hackers, viruses, worms, and even emerging threats like ransomware attacks. It involves staying abreast of the latest developments in cybersecurity and evolving threat actor tactics.

Once the risks are identified and analysed, they must be evaluated. This involves estimating the likelihood of a risk occurring and its potential impact. This process is often subjective, depending on the nature of the organisation, its industry, its digital infrastructure, and its risk tolerance.

A significant part of this chapter will be dedicated to discussing how to conduct risk assessments, including various methodologies, tools, and techniques used in the process. From qualitative and quantitative methods to automated tools, we will explore different ways to effectively conduct risk assessments.

A thorough risk assessment is a cornerstone of any cybersecurity program. It offers an accurate understanding of the organisation's risk profile, which informs decision-making, guides resource allocation, and shapes the entire cybersecurity strategy. With a comprehensive risk assessment, organisations can make informed decisions and develop robust strategies to manage cyber risks, fostering resilience and ensuring business continuity.

9.1.1. Steps in Performing a Cybersecurity Risk Assessment

A cybersecurity risk assessment is a systematic process, involving careful examination of an organisation's digital environment to identify potential vulnerabilities and threats, and to evaluate the potential impact should those vulnerabilities be exploited. It forms the foundation upon which effective cybersecurity strategies are built, serving as a roadmap for organisations to navigate the complex landscape of cybersecurity. The following are the essential steps involved in performing a successful cybersecurity risk assessment.

1. **Identify Assets:** At the core of a risk assessment is the identification of an organisation's assets. These assets could range from hardware devices and software applications to databases and networks, and even include intangible assets like data and reputation. Asset identification is an important precursor to understanding the various points of vulnerability that a threat could potentially exploit. It's not just about listing assets, but also about understanding their value, their function in the broader system, and the data they handle or process.

2. **Identify Threats:** With a comprehensive list of assets at hand, the next step is to identify potential threats. This is a broad step that covers a vast terrain of possibilities. Threats could stem from natural disasters that could physically damage the infrastructure, to human error, to deliberate attacks from malicious hackers. An understanding of the threat landscape involves staying updated with the latest types of cyberattacks, attacker tactics, techniques, and procedures.

3. **Identify Vulnerabilities:** Once threats are identified, the next step is to look inward and identify vulnerabilities within the organisation's system. Vulnerabilities could be inherent in the design of a software or hardware component, or they could be the result of improper configurations, outdated software, weak passwords, or even lapses in user awareness. This step often involves a combination of manual review and automated vulnerability scanning tools.

4. **Analyse Risk:** Once threats and vulnerabilities are identified, they need to be paired to analyse the potential risk. This involves estimating the likelihood that a given threat would exploit a particular vulnerability, and the potential impact if it does. This

analysis should also consider the current controls in place to mitigate these risks.

5. **Prioritise Risks:** With the risk landscape charted, organisations must then prioritise these risks. Not all risks carry the same weight. Some risks may have a high likelihood of occurrence but low impact, while others may be less likely but carry catastrophic consequences. Prioritisation should consider the organisation's risk appetite and should align with its strategic goals and regulatory requirements.

6. **Develop a Risk Management Strategy:** With a clear understanding of the risk landscape and the prioritised risks, the final step is to develop a risk management strategy. This strategy should articulate how the organisation intends to treat each risk – whether they plan to mitigate it (by implementing controls), transfer it (through insurance, for example), avoid it (by changing business processes or technology), or accept it (if the cost of treatment outweighs the potential impact).

This process of conducting a cybersecurity risk assessment is not a one-off event. It's a continual process that needs to be repeated at regular intervals and whenever significant changes occur within the organisation's systems or in the external threat landscape. A regular risk assessment allows organisations to stay ahead of potential threats and respond proactively, enhancing their cybersecurity posture in an ever-evolving cyber environment.

9.1.2. Risk Assessment Tools and Methodologies

Cybersecurity risk assessments are crucial in maintaining an effective defence against an array of potential threats. Yet, the task of conducting such an assessment can be daunting, particularly given the complexity and scale of today's digital environments. Luckily, there exist a variety of tools and methodologies to aid this process, streamlining and standardising the approach to risk assessment, and ensuring a thorough and consistent output.

Risk Assessment Tools

Tools play a crucial role in risk assessments by automating many aspects of the process and increasing the efficiency of the assessment. These tools can help identify vulnerabilities, quantify risks, and provide risk mitigation recommendations.

1. **Vulnerability Scanners:** Tools like Nessus or OpenVAS scan a network for known vulnerabilities, which can help an organisation to identify weak points in their systems and infrastructure that can be exploited by attackers.

2. **Security Information and Event Management (SIEM) Systems:** SIEM systems, such as Splunk or LogRhythm, aggregate and analyse log and event data from various sources within an organisation, helping to identify anomalies and potential threats.

3. **Risk Quantification Tools:** These tools, like RiskLens or RiskWatch, facilitate the process of risk quantification, helping to measure the potential impact of a given risk on an organisation's assets.

Risk Assessment Methodologies

There are several well-recognised methodologies that provide a structured approach to risk assessment.

1. **NIST SP 800-30:** Developed by the National Institute of Standards and Technology (NIST), this guide provides a step-by-step approach to conducting a risk assessment, including identifying threats, vulnerabilities, impacts, and risk responses.

2. **ISO 27005:** As part of the ISO 27000 series of standards on information security management, ISO 27005 provides guidelines for information security risk management, including principles, a framework, and a process for risk assessment.

3. **OCTAVE (Operationally Critical Threat, Asset, and Vulnerability Evaluation):** OCTAVE is a suite of tools, techniques, and methods for risk-based information security strategic assessment and planning. It involves the identification of

organisational objectives and vulnerabilities and the development of a risk mitigation strategy.

4. **FAIR (Factor Analysis of Information Risk):** FAIR is a quantitative risk analysis methodology that helps organisations understand, analyse, and quantify information risk in financial terms.

These tools and methodologies aid organisations in conducting thorough, repeatable risk assessments. By selecting the appropriate tools and methodologies that fit their needs, organisations can ensure they have a detailed understanding of their risk landscape and are well-prepared to mitigate those risks effectively. Regardless of the specific tools or methodologies used, it's crucial that organisations view risk assessment as an ongoing process, reflecting the dynamic nature of both threats and the organisation's own information systems.

9.1.3. Interpreting and Acting on Risk Assessment Results

Upon completing a cybersecurity risk assessment, organisations are faced with the critical task of interpreting the results and deciding upon an appropriate course of action. These findings, when correctly interpreted and acted upon, can transform an organisation's cybersecurity posture, enhance decision-making, and contribute to a more resilient and proactive security strategy.

Interpreting Risk Assessment Results

Interpreting the results of a risk assessment involves understanding the identified vulnerabilities, the associated threats, the potential impact, and the probability of occurrence. Key factors to consider in this interpretation include:

1. **Risk Score:** Many risk assessment methodologies and tools provide a risk score based on factors such as the potential impact of a risk and the likelihood of its occurrence. Higher scores typically indicate higher risks.

2. **Risk Classification:** Results often categorise risks into classifications such as low, medium, and high. This can help organisations prioritise their response and resources.

3. **Risk Impact:** Understanding the potential impact of a risk on an organisation's operations, reputation, and bottom line is critical in interpreting risk assessment results. This might involve quantifying the potential financial loss or qualitatively assessing the damage to reputation or operational disruption.

4. **Vulnerability Details:** Detailed information on identified vulnerabilities can provide insights into weak points in the system or process and potential avenues for threat actors to exploit.

Acting on Risk Assessment Results

Once risk assessment results are interpreted, organisations need to decide how to act on these findings. The chosen course of action will depend on the risk appetite of the organisation, the potential impact of the identified risk, and the resources available for risk mitigation. Possible courses of action include:

1. **Risk Mitigation:** This involves implementing controls to reduce the risk to an acceptable level. Controls can be administrative (policies, procedures), technical (firewalls, encryption), or physical (locks, biometrics).

2. **Risk Transfer:** In some cases, risks might be transferred to another party, such as through insurance or outsourcing.

3. **Risk Acceptance:** If a risk is deemed acceptable given its potential impact and the cost of mitigation, it may be accepted. This decision should be informed and intentional.

4. **Risk Avoidance:** If a risk is too great, and mitigation is not feasible or cost-effective, an organisation may choose to avoid the risk altogether, perhaps by discontinuing the associated service or process.

All these actions should be documented in a risk management plan, which outlines the identified risks, the chosen course of action for each, and the parties responsible for implementing those actions. In this way, risk assessment results become a catalyst for proactive cybersecurity measures, guiding strategic decisions and enabling organisations to navigate the complex cybersecurity landscape with a comprehensive, risk-informed approach.

9.2. Developing a Cybersecurity Policy

An organisation's cybersecurity policy is the cornerstone of its security posture, laying out the rules, expectations, and guidelines that govern the protection of its digital environment. The policy itself is a potent instrument that conveys the principles and values of an organisation's approach to cybersecurity. It underpins the culture of security within an organisation, providing a solid structure upon which all other initiatives are built.

This section will traverse the methodological and philosophical terrain of crafting such a policy. It will walk you through the integral steps, from identifying the purpose and scope of the policy to detailing the specific roles and responsibilities. Recognising that each organisation's cybersecurity needs are unique, the chapter underscores the need for an in-depth risk assessment and identification of critical assets, allowing the policy to align closely with the specific challenges and goals of the organisation.

The chapter underscores the pivotal role of identifying and implementing the appropriate security measures and controls, illustrating how these elements collectively form the defensive wall around an organisation's digital assets. Delving further, the discussion will expand to the crucial concepts of incident response and disaster recovery procedures, highlighting the importance of preparedness and resilience in the face of inevitable cybersecurity incidents.

Moreover, this chapter encourages the idea of cybersecurity policy as a dynamic and evolving document. With the rapidly changing cybersecurity landscape, the policy needs regular review and update, allowing it to remain relevant and effective amidst new threats and vulnerabilities. The process of policy development is continuous and iterative, necessitating active engagement and vigilance.

By the end of this section, you'll have a deep understanding of the significant elements that make up a robust cybersecurity policy, equipped with the knowledge to create or enhance your organisation's approach to information security. This chapter aims to empower you with the insights needed to navigate the complex waters of cybersecurity policymaking with competence and confidence. It provides the steppingstone towards fostering a security-conscious culture within your organisation, one where every stakeholder understands their role in safeguarding the digital assets.

9.2.1. Key Elements of a Comprehensive Cybersecurity Policy

A cybersecurity policy serves as the compass guiding an organisation's digital journey. It offers a blueprint detailing strategic actions to safeguard data and systems, reinforcing their resilience against adversarial threats. However, it's important to note that drafting an effective cybersecurity policy requires a clear understanding of its key elements. Here, we delve into these integral parts, casting light on their roles in establishing a solid cybersecurity foundation.

First, the purpose statement is essential, setting the stage for the entire policy. It outlines the rationale behind the policy's implementation, explaining its relevance to the organisation's mission and operations. This statement doesn't merely chronicle the policy's intent but also resonates with stakeholders, instilling a shared understanding of the importance of information security.

Second, the scope of the cybersecurity policy must be defined. The scope identifies which domains of the organisation the policy applies to, covering aspects such as departments, personnel, systems, and data types. Defining a clear scope ensures that no critical area is overlooked, and the expectations and responsibilities of each entity under the policy's purview are clearly set.

The next element is the clear definition of roles and responsibilities. This section articulates who is accountable for implementing, enforcing, and maintaining the policy. It defines the key roles within the cybersecurity structure, including system users, system administrators, security officers, and top management, outlining what is expected from each in terms of maintaining and enforcing security measures.

The cybersecurity policy must also establish rules for acceptable use of systems and data. These rules define how users are allowed to access and interact with the organisation's digital assets, from hardware to software and data. The acceptable use policy (AUP) is particularly important as it sets forth the do's and don'ts for system users, highlighting the need to avoid risky behaviours such as visiting insecure websites or downloading unauthorised software.

Equally important is the inclusion of incident response procedures. Given that cybersecurity incidents are a matter of 'when', not 'if', the policy should delineate the protocols to follow when an incident occurs. This includes steps to detect, contain, eradicate, and recover from security breaches, and a clear escalation process.

The policy should also highlight the organisation's approach to training and awareness. After all, employees are the first line of defence against cyber threats. Therefore, regular training programs to keep them abreast of the latest threats and safe practices, as well as continuous awareness campaigns to keep cybersecurity at the forefront of their minds, are essential.

Lastly, the cybersecurity policy should outline the review and update processes. As the cyber threat landscape evolves rapidly, the policy should not remain static. It must be regularly reviewed and updated to remain relevant in the face of emerging threats and technological advancements. A review schedule should be established, including triggers for review outside of this schedule, such as significant changes in operations or security incidents.

In summary, a comprehensive cybersecurity policy is a multi-layered construct that demands thorough understanding and careful planning. Its various elements interlock to form the protective bulwark that guards an organisation's digital realm, underscoring the policy's role as an instrument of proactive defence. The key lies in creating a policy that is both robust and flexible, capable of withstanding the myriad challenges of the current threat landscape while adapting to its future evolutions.

9.2.2. Writing and Implementing a Cybersecurity Policy

The process of writing and implementing a cybersecurity policy demands a meticulous approach, underpinned by a profound understanding of the organisation's goals, assets, and potential threats. In this endeavour, the importance of clarity, accuracy, and usability cannot be overemphasised. Successful drafting and execution of a cybersecurity policy play a pivotal role in solidifying an organisation's defences, paving the way for a resilient and secure digital environment.

The initial step in writing a cybersecurity policy involves a thorough assessment of the organisation's cyber landscape. It entails identifying the digital assets that require protection, from hardware and software to confidential data, and understanding the potential threats they face. This assessment aids in crafting a policy that is contextually relevant and practically effective.

An effective cybersecurity policy is a detailed document that offers explicit guidance. When writing the policy, ensure that it communicates the organisation's cybersecurity goals and strategies clearly. Each section of the policy should be comprehensible to all stakeholders, avoiding jargon that could lead to misunderstanding or ambiguity. Remember, a policy that is clear and comprehensible fosters greater adherence and consistency in security practices.

Inclusion is another crucial aspect when writing a cybersecurity policy. The document should cover all facets of cybersecurity relevant to the organisation, including physical security, access control, data privacy, incident response, and user behaviour, among others. Additionally, it should consider various scenarios, ranging from everyday operations to emergency situations, ensuring the organisation is prepared for a multitude of cybersecurity events.

Once the policy is written, attention shifts to its successful implementation, which is a multi-staged process. Firstly, top-level management must endorse the policy, exhibiting their commitment to cybersecurity. This step is critical as it ensures that the policy receives the necessary support and resources for its execution.

Next, disseminate the policy across the organisation. Every member of the organisation should be aware of the cybersecurity policy and understand its implications. Communication methods can vary from emails, policy handbooks, or a dedicated section on the organisation's intranet.

Training and education form the cornerstone of successful policy implementation. Conduct comprehensive training programs that detail the policy's provisions and the roles and responsibilities of each individual. This not only improves the employees' understanding of the policy but also reinforces its importance in maintaining the organisation's cybersecurity.

Another critical step in policy implementation involves integrating the policy with existing systems and procedures. This may require updates or changes to system configurations, implementation of new security tools, or modifications to access controls.

Lastly, regular monitoring and enforcement are crucial for ensuring the policy's effectiveness. Regular audits and compliance checks should be conducted to determine if the policy is being adhered to and identify areas where improvement is needed. Any violations should be dealt with promptly, with appropriate consequences as defined in the policy.

In conclusion, writing and implementing a cybersecurity policy is a complex yet rewarding endeavour. It not only demands comprehensive understanding of the organisation's cybersecurity landscape and meticulous planning but also entails effective communication and consistent enforcement. The end result, however, is a robust cybersecurity posture that significantly enhances the organisation's resilience against cyber threats.

9.2.3. Periodic Review and Update of the Cybersecurity Policy

In the ever-changing panorama of cyber threats and digital innovation, a static cybersecurity policy is as obsolete as yesterday's newspaper. As digital threats morph with alarming speed, technological advancements surge ahead relentlessly, and business dynamics shift like desert sands, it is the mandate of every cybersecurity policy to keep pace. Therefore, the consistent review and updating of a cybersecurity policy is not merely an optional extra, but a core component of a viable cybersecurity strategy.

Periodic evaluation of a cybersecurity policy encompasses a thorough re-examination of its elements to validate its continued relevance to organisational goals and current threat scenarios. This critical process can shed light on inadequacies or redundancies within the policy, underscore emerging threat patterns, and reveal opportunities for augmentation, thus ensuring optimal policy effectiveness.

The initial stage of the review process entails a comprehensive analysis of the organisation's current cybersecurity standing and an assessment of the existing policy's efficacy. This involves dissecting past security incidents, understanding their underlying causes, examining the response mechanism, and evaluating whether the policy provided adequate and accurate guidance. Furthermore, the analysis should take into account any shifts in the organisation's structure, strategy, technology stack, and regulatory backdrop that might demand policy modifications.

Following the review, the process of updating the cybersecurity policy is driven by the intelligence gathered. It might involve minor tweaks, such as refining protocols or definitions, or necessitate significant overhauls, such as crafting new sections to address emergent threats or technologies. The updated policy should be clear, exhaustive, and flexible, mirroring the fluid nature of the digital world.

However, even the most impeccably updated policy is worthless if it fails to permeate the organisation. Thus, the updated policy must be effectively communicated across the company, ensuring that all employees are cognisant of the revisions. Furthermore, any necessary alterations in procedures, system settings, or tools, based on the updated policy, should be executed without delay.

The frequency of policy reviews and updates is another critical variable. While there is no universally applicable timeline, it is recommended that cybersecurity policies be reviewed at a minimum on an annual basis. However, in light of significant changes within the organisation or in the cyber threat landscape, more frequent reviews could be necessitated.

In summary, the iterative process of regularly reviewing and updating a cybersecurity policy is indispensable for the cybersecurity robustness of an organisation. This continual cycle ensures the policy stays pertinent and robust, fostering an adaptive cybersecurity culture that evolves synchronously with the broader digital milieu.

9.3. Implementing Security Controls

Diving into the heart of cybersecurity, we land on the subject of implementing security controls. These controls are the protective measures and actions taken to shield an organisation's digital assets from a myriad of threats. Security controls not only offer protection from external threats but also internally—whether they are intentional malicious acts or accidental oversights by employees.

In the 21st-century enterprise, data is the primary asset—making it the most sought after by adversaries. Thus, effective and robust security controls are the cornerstone of an organisation's cybersecurity infrastructure. They function as the building blocks that support and maintain the integrity, confidentiality, and availability of the systems and data—forming the core of every cybersecurity program.

The role of security controls expands beyond merely safeguarding an organisation's systems and data. They contribute significantly towards compliance with regulatory requirements, industry standards, and best practices—making their effective implementation non-negotiable.

The domain of security controls spans across several areas—from technical implementations such as firewalls, encryption, and intrusion detection systems, to administrative controls such as policies and procedures, to physical controls that restrict access to critical resources.

The implementation of security controls involves careful planning, execution, and monitoring to ensure their effectiveness. But, before this process begins, it is important to understand the specific risks to the organisation's information systems. Risk assessment plays a crucial role in identifying these risks and defining the appropriate controls to mitigate them.

In the upcoming sections, we will delve deeper into the specifics of implementing security controls, exploring the different types and their relevance in ensuring a robust cybersecurity infrastructure. Understanding these controls and their application will enable organisations to enhance their security posture, safeguard their assets, and maintain the trust of their stakeholders.

9.3.1. Choosing the Right Security Controls for Your Organisation

Choosing the right security controls for your organisation isn't just a matter of selecting the most expensive tools or copying what successful companies are doing. It's a process that involves thorough understanding of your specific circumstances, assessing your organisation's risk landscape, identifying vulnerabilities, and then adopting the most appropriate set of controls that meet your needs.

The journey starts with comprehending your organisation's mission, vision, and strategic objectives. What are your critical assets and business processes? What level of risk is your organisation ready to accept? Understanding this information is crucial to align security controls with business objectives and to ensure that your security strategy supports your mission rather than hindering it.

Next, a comprehensive risk assessment should be conducted. This involves identifying all potential threats and vulnerabilities that your organisation faces, from cyber-attacks to physical security threats. It is also important to consider the probability of these risks materialising and their potential impact. This step is critical because it forms the basis for the selection and prioritisation of security controls.

Once the risks are identified, the focus shifts to understanding what controls are available to mitigate these risks. Security controls fall into three categories: technical controls (like encryption, antivirus software, and firewalls), administrative controls (like security policies and procedures, training, and awareness programs), and physical controls (like locks, access cards, and surveillance cameras).

Your choice should also be driven by the legal, regulatory, and industry-specific compliance requirements your organisation needs to meet. From PCI DSS in the payment card industry to GDPR for personal data protection, different industries face different regulations, and non-compliance can result in hefty fines and reputational damage.

Another significant consideration is the cost—both direct and indirect. While direct costs involve the price of acquiring and implementing the control, indirect costs may include staff training, maintenance, and the potential impact on business operations. The cost must be balanced with the benefits of risk reduction.

Remember, no security control is fool proof, and an approach that combines various types of controls will often provide a stronger defence. It's also vital to understand that security is a dynamic field, and what works today may not work tomorrow. Therefore, regular review and updating of controls are necessary to keep up with evolving threats and business changes.

Lastly, the right security controls are only as effective as the organisation's commitment to implementing and maintaining them. Executive buy-in is crucial for a successful cybersecurity program. Every member of the organisation, from top to bottom, should understand their role in maintaining the organisation's security posture.

Choosing the right security controls requires strategic thinking, a thorough understanding of your organisation, and a commitment to cybersecurity. The stakes are high, but with the right approach, you can create a secure environment that protects your assets, ensures business continuity, and builds trust among stakeholders.

9.3.2. Implementing and Monitoring Security Controls

Implementing and monitoring security controls is a process that extends beyond the mere installation of security tools and technologies. It requires a holistic approach that encompasses strategy, planning, training, execution, and ongoing review. With the ever-evolving cybersecurity landscape, today's organisations must ensure they are consistently vigilant and proactive in their approach to security.

Begin by strategising and planning. This includes the creation of detailed implementation plans that identify necessary resources, define roles and responsibilities, and establish timelines. Be cognisant of the need for expertise and skill sets necessary to implement each control. In some cases, this might mean investing in training or even bringing in external experts.

As controls are implemented, it is vital to document all changes to your systems and configurations. Documentation will be a significant aid when troubleshooting issues, providing evidence during audits, and helping maintain a stable and secure environment.

Training is another crucial element. Security controls aren't just about technology; they're also about people and processes. Hence, all employees need to be educated about their role in maintaining these controls, what they are intended to do, and how to use them effectively.

Implementation is only the first half of the battle. Once security controls are in place, they must be continually monitored to ensure they're functioning as intended and protecting your organisation against threats. This ongoing process helps identify anomalies, intrusions, and any other signs of potential breaches. Monitoring can range from automated system logs to more complex security information and event management (SIEM) systems that consolidate and analyse data from various sources.

Keep in mind that the work is not done once controls are implemented and monitored. As your organisation evolves – in size, scope, or operations – so too will its vulnerabilities. Regular audits and assessments are therefore vital to identifying gaps in your security posture. Use the insights gained from these evaluations to revise and update your controls.

Implementing and monitoring security controls is an ongoing journey. The landscape of threats and vulnerabilities is continually changing, and as such, organisations must be diligent and proactive in maintaining and updating their security controls. With the right blend of strategy, execution, and continuous review, your organisation can navigate the dynamic field of cybersecurity, ensuring not just survival, but success in today's interconnected world.

9.3.3. Evaluating the Effectiveness of Security Controls

The process of evaluating the effectiveness of security controls is a fundamental element of a robust cybersecurity program. It's through this evaluation that organisations can understand how well their security mechanisms are performing and where improvements are needed. This process isn't a one-time event; instead, it should be conducted regularly and systematically to ensure that controls remain effective in the face of evolving threats and changing business operations.

The process starts with defining metrics for the effectiveness of each control. These metrics, often referred to as Key Performance Indicators (KPIs), should be tailored to each control and may be qualitative or quantitative. For example, the effectiveness of an intrusion detection system might be measured by the number of detected and prevented attacks, while user awareness training might be evaluated based on the results of phishing simulation tests.

Automated tools, such as Security Information and Event Management (SIEM) systems, can provide valuable data for evaluating controls. SIEM systems gather logs and other data from throughout the network, allowing for real-time analysis and reporting. This automated data collection can reveal trends, anomalies, or weak points that may be overlooked in manual evaluations.

It's important to recognise that the effectiveness of controls can change over time. As new vulnerabilities are discovered and new types of attacks are developed, controls that were once effective may become less so. Regular reviews should be scheduled to assess and adjust controls as necessary.

Another significant part of evaluating control effectiveness is penetration testing and vulnerability scanning. These methods simulate attacks on the organisation's systems to identify potential weaknesses before they can be exploited by an attacker. They can provide practical, real-world data on how well controls are performing.

While automated tools and systems play a vital role in evaluation, human insight is equally important. Regular meetings with IT staff, security teams, and other stakeholders can provide invaluable insights into the effectiveness of controls. Their experience on the front lines of defence can identify subtle issues that automated systems might miss.

Lastly, it's crucial to document the results of the evaluations and share them with key stakeholders. This documentation provides a historical record that can help identify trends, demonstrate compliance with regulatory requirements, and inform decision-making about future investments in security.

In summary, evaluating the effectiveness of security controls is a critical process that helps organisations maintain a strong security posture. By defining appropriate metrics, utilising automated tools, conducting regular reviews, and leveraging human insights, organisations can ensure that their controls remain effective in the face of a continually evolving threat landscape.

9.4. Incident Response Planning and Execution

In the world of cybersecurity, there is a maxim that resonates through the boardrooms and server rooms alike: It's not a matter of if, but when. The landscape of digital threats is vast and complex, underscoring the importance of an effective Incident Response Plan (IRP) to an organisation's cybersecurity strategy. A well-executed IRP is the difference between a minor disruption and a disastrous event; it can significantly reduce the duration and impact of a security incident, saving resources, protecting reputation, and ultimately ensuring business continuity.

Incident response planning is an iterative process designed to handle security incidents in a controlled and systematic way. It involves identifying potential incidents, establishing protocols for addressing them, and assigning roles and responsibilities within the organisation. The intent is to manage the situation in a way that limits damage, reduces recovery time and costs, and safeguards pertinent information from falling into the wrong hands.

As we delve deeper into this section, we will explore various elements that constitute a robust incident response plan. We will walk you through the process of incident identification, containment strategies, eradication and recovery mechanisms, and post-incident analysis to learn and adapt. Additionally, we will provide insights into incident response team structures, their critical roles, and the necessary tools and technologies that aid in timely and effective incident resolution.

Remember, the ultimate goal of an incident response plan is not just to react to incidents when they occur but to prepare the organisation to handle incidents efficiently and effectively, minimising impact and downtime. As we move forward, we'll provide a roadmap for creating, implementing, and refining your incident response plan to best suit your organisation's unique needs and capabilities. Let's begin this journey towards strengthening your organisational resilience in the face of cybersecurity incidents.

9.4.1. Developing an Incident Response Plan

Creating an Incident Response Plan (IRP) is a critical undertaking that demands a comprehensive and strategic approach. The plan's objective is clear: to provide a detailed roadmap that your organisation can follow when a security incident occurs. From minor glitches to significant breaches, your IRP will ensure your organisation is prepared to address any scenario quickly and efficiently.

Your first consideration when developing an IRP is understanding the threats your organisation faces. Are you more susceptible to insider threats or external attacks? What types of data are you storing and processing? What are the consequences if that data is compromised? Answering these questions will help you identify the most likely risks and focus your response planning accordingly.

Once you understand your risk landscape, start by defining what constitutes an incident for your organisation. It could be a successful phishing attempt, malware detection, unauthorised access, or a variety of other situations. Be specific and inclusive, ensuring that everyone within the organisation can identify potential incidents.

Next, your IRP needs to have a clear incident classification system. Not all incidents are created equal; some are mere annoyances, while others pose significant threats to your operations or reputation. Developing a tiered classification system will help prioritise response efforts and allocate resources where they are needed most.

The heart of your IRP is the response process. This process should be broken down into stages such as detection, analysis, containment, eradication, and recovery. The detection phase involves identifying the incident, while the analysis phase entails understanding the nature and extent of the incident. Containment is all about limiting the spread of the incident, whereas eradication aims at eliminating the root cause. The recovery phase ensures that affected systems and data are restored to their pre-incident state, and operations return to normal.

Each stage should include detailed procedures for action, with clearly defined roles and responsibilities. It's essential that all relevant stakeholders know their tasks and whom to report to, ensuring a swift and coordinated response. Be sure to provide comprehensive training so everyone involved understands their role in the plan.

A valuable aspect of your IRP is communication. Your plan should outline communication protocols for keeping internal stakeholders informed during an incident and the chain of command. If necessary, it should also include guidelines for external communication, such as notifying affected customers, partners, or regulators.

Lastly, your IRP should be a living document, not a static one. Threats and vulnerabilities evolve; therefore, your incident response plan must as well. Incorporate a regular review and update process into your plan. After every incident, conduct a post-incident review to learn from the event and improve your response in the future.

Remember, the aim of developing an Incident Response Plan is not to create an infallible shield against cyber threats - no such thing exists. Instead, its purpose is to reduce the chaos and confusion that often accompany security incidents, enabling your organisation to react swiftly and decisively. It's about minimising damage, preserving trust, and ensuring business continuity in the face of a cyber incident. The result of this effort is an organisation that is more resilient, prepared, and confident in its ability to navigate the unpredictable seas of the digital world.

9.4.2. Incident Response Team: Roles and Responsibilities

An effective Incident Response Plan (IRP) heavily relies on the Incident Response Team (IRT), a group of individuals tasked with implementing the plan. Each member has specific roles and responsibilities, ensuring the team works together in a coordinated and efficient manner to mitigate and resolve security incidents.

At the helm of the IRT is the Incident Response Manager. This individual oversees the entire incident response process, from the initial detection to the final recovery. Their primary role is to ensure that all incidents are handled according to the established procedures, and they must make critical decisions during the incident, often under significant pressure. Their responsibilities also include communicating with senior management, coordinating with other teams, and managing any third-party involvement.

Technical experts form the backbone of the IRT. These are IT professionals with specialised skills in various areas such as network security, forensics, malware analysis, and system administration. Their role is to investigate the incident, understand its nature and scope, and apply their expertise to contain, eradicate, and recover from the incident. It's crucial for these experts to keep up to date with the latest threats, vulnerabilities, and response techniques.

The role of the Legal Advisor within the IRT is to ensure that all actions taken during the incident response process comply with relevant laws and regulations. They also advise on potential legal ramifications, such as liability issues and reporting obligations. In some cases, they might work with external legal counsel, especially when dealing with severe incidents that could lead to litigation or regulatory action.

The Public Relations (PR) Officer is tasked with managing external communications during and after a security incident. They prepare statements for customers, the media, and other stakeholders, aiming to maintain the organisation's reputation. It's important for the PR Officer to strike a balance between transparency and discretion, sharing enough information to maintain trust without revealing sensitive details that could exacerbate the situation or give away tactical advantage to the adversaries.

An IRT also often includes a representative from the Human Resources (HR) department, especially in incidents involving internal actors. The HR representative handles any people-related matters that arise, such as disciplinary actions, potential dismissals, or counselling for affected employees.

Lastly, every member of the IRT should be an advocate for lessons learned. After an incident is resolved, the team should conduct a post-incident review to analyse what happened, identify what worked and what didn't, and suggest improvements to the IRP. This ongoing learning and adaptation are critical for building resilience against future incidents.

In conclusion, an effective IRT isn't just about having a diverse set of skills. It's also about effective communication, collaboration, and coordination among team members. With a well-defined and well-functioning IRT, your organisation can respond to cyber incidents in a way that minimises damage, ensures quick recovery, and maintains trust with customers and partners.

9.4.3. Incident Response: Detection, Analysis, Containment, Eradication, and Recovery

Incident response is a strategic approach to handling security incidents, cyber threats, and breaches. It involves several essential steps: Detection, Analysis, Containment, Eradication, and Recovery. Let's delve deeper into each of these steps, demonstrating their significance and execution strategies.

Detection is the first step in the incident response process. It involves the identification of potential security incidents. This can be achieved through various means, including the use of intrusion detection systems (IDS), security information and event management (SIEM) solutions, user and entity behaviour analytics (UEBA), and other monitoring tools. Alerting mechanisms should be in place to inform the relevant team members once an incident is detected.

The second step, Analysis, involves investigating and validating the detected incident. This requires a thorough examination of the data available, which can include log data, network traffic, user activity, and more. The aim is to understand the nature of the incident: the systems, data, or resources affected, and the potential impact. In some cases, threat intelligence feeds can be used to provide context and inform the analysis.

Once the incident has been verified and understood, the next step is Containment. The objective here is to limit the impact of the incident and prevent it from spreading further. Depending on the nature of the incident, containment strategies can vary. They might involve disconnecting affected systems from the network, blocking specific IP addresses, or changing user credentials.

Eradication is the next stage, which aims at eliminating the cause of the incident. It often involves the removal of malicious code, unauthorised users, or any components that enabled the breach. Depending on the incident's complexity, eradication can be quite challenging and may require significant technical expertise. It's also important to identify and address the vulnerabilities that allowed the incident to occur, to prevent recurrence.

The final step is Recovery, where the affected systems are restored and returned to their normal operations. It may involve patching software, restoring data from backups, verifying system integrity, and confirming that all traces of the incident have been removed. The recovery phase is also when you start to consider the long-term implications of the incident and start planning for the future to prevent similar incidents.

An essential part of the incident response process is communication. All stakeholders should be kept informed about the situation, the actions being taken, and their roles in the process. Moreover, once the incident has been resolved, a post-incident review should be carried out. The aim here is to learn from the incident and improve the incident response process based on those learnings.

In essence, an effective incident response plan involves a structured approach for handling cyber threats, which can minimise the impact of an incident and ensure swift recovery. Each stage plays a critical role and is indispensable for the plan's overall success. By executing these steps efficiently, organisations can enhance their cybersecurity posture and resilience against future incidents.

9.5. Staff Training and Awareness Programs

As we venture into Section 9.5, we shift our focus towards the human element of cybersecurity – Staff Training and Awareness Programs. This topic gains importance because cybersecurity is not just a matter of technology and processes, but also of people. Human errors, misunderstandings, or lack of awareness can introduce significant vulnerabilities into an otherwise secure system. Hence, it is crucial to equip employees with the right understanding and skills to navigate the digital landscape securely.

In this section, we'll discuss how developing a comprehensive training program can boost an organisation's security posture. We'll explore the methods of communicating the importance of cybersecurity, teaching best practices, and creating a culture of security. We'll also highlight how to tailor these programs according to the needs of different roles and levels within the organisation. Additionally, we'll underscore the need for ongoing training and awareness activities to stay up to date with the evolving cybersecurity landscape.

Remember, the goal is not to make every employee a cybersecurity expert but to cultivate an environment where security is valued, and everyone contributes to it. A robust staff training and awareness program is a pivotal part of any organisation's cybersecurity strategy. Let's delve deeper into the particulars of creating such a program and its substantial impact on reinforcing an organisation's security defences.

9.5.1. Importance of Cybersecurity Awareness Training

In the contemporary digital ecosystem, an organisation's cybersecurity stance relies significantly on its employees' awareness. Cybersecurity awareness training is not merely a nice-to-have; it is an imperative strategy in the fight against cyber threats. By understanding the importance of this training, organisations can make a robust case for investing in these initiatives.

The cybersecurity landscape is a constantly evolving battlefield, where new threats emerge every day. The tactics used by cybercriminals are increasingly sophisticated, with many focusing on exploiting human error. Phishing attacks, social engineering, ransomware, and advanced persistent threats are just a few of the many ways cybercriminals can penetrate an organisation's defences.

The people within an organisation are both its greatest asset and potentially its biggest weakness in terms of cybersecurity. Humans can be prone to errors, and a simple mistake, such as clicking on a malicious link or using a weak password, can have significant consequences. According to a report by Verizon, nearly a third of all data breaches involve some form of human error. Thus, ensuring that employees are aware of the potential threats they face and how to mitigate these risks is crucial.

Cybersecurity awareness training is designed to educate staff about the nature of these threats and how they can protect themselves and the organisation. It encompasses a range of areas, from understanding the signs of phishing emails and malware-infected attachments to the importance of strong passwords and following security procedures.

Training programs also highlight the potential consequences of security breaches. When employees comprehend the real-world implications of a cyber-attack, such as financial loss, reputational damage, and regulatory fines, they are more likely to take their role in preventing these incidents seriously.

Importantly, cybersecurity awareness training isn't only about preventing attacks. It also involves educating employees on what steps to take in case a breach occurs. Quick action can help limit damage, isolate the problem, and speed up recovery times.

Moreover, training plays a critical role in maintaining compliance with various data protection regulations, such as the GDPR or HIPAA. Many of these regulations require organisations to provide their employees with training on handling personal data securely. Thus, a comprehensive training program helps organisations meet their legal obligations and avoid hefty fines.

Lastly, cybersecurity awareness training contributes to building a strong security culture within the organisation. It fosters a proactive attitude towards cybersecurity, encouraging employees to be vigilant and mindful of their actions. It empowers them to take responsibility for their role in maintaining the organisation's security posture.

To conclude, cybersecurity awareness training is a cornerstone of effective cybersecurity strategy. It helps protect against the increasing number of threats, reduces the likelihood of human errors leading to security breaches, enhances compliance with data protection regulations, and cultivates a robust security culture within the organisation. Thus, investing in regular and comprehensive cybersecurity awareness training is not just beneficial, but indispensable in today's cyber threat landscape.

9.5.2. Designing and Implementing a Cybersecurity Training Program

Crafting and instituting a cybersecurity training program is no small undertaking. It demands a balanced blend of comprehensiveness, inclusivity, and adaptability, all underpinned by an empathetic understanding of the human element. To create a successful cybersecurity training program, it's important to remember that it's not about developing an impenetrable human firewall but about fostering a community of informed and conscious custodians of digital assets.

At the outset, it's essential to identify the unique needs, risks, and challenges that your organisation faces. Every enterprise has a distinctive digital footprint, and thus a unique vulnerability profile. Understanding this profile is the first step in designing a tailored training program. This process often involves surveys, interviews, or consultations with various stakeholders within the organisation.

Once you have a clear picture of your organisation's needs, you can begin developing the content of the training. The material should cover a wide range of topics, from basic cyber hygiene practices such as creating strong passwords and recognising phishing attempts, to more advanced topics, including the handling of sensitive data and adherence to industry-specific security regulations.

Although it's essential to provide comprehensive information, it's equally crucial to deliver it in an engaging and accessible way. Lengthy presentations packed with technical jargon are likely to be forgotten or misunderstood. Instead, consider employing a variety of learning modalities such as interactive activities, gamification, or scenario-based learning. This variety can not only make the training more engaging but also accommodate different learning styles.

The next critical step is to ensure that the training reaches all members of the organisation. Cybersecurity is not just an IT issue but an organisational one. Everyone, from the C-suite to the frontline employees, should be involved in the training. Remember, a chain is only as strong as its weakest link, and in the case of cybersecurity, any employee could potentially be that link.

Once the program is in place, don't consider the job done. The dynamic nature of the cybersecurity landscape requires constant updates and modifications to the training content. Regularly review and update your training material to keep up with the evolving threat landscape. In addition, regular evaluations or assessments can help measure the effectiveness of the training and identify areas for improvement.

Lastly, make sure to foster an environment where questions are welcomed, and continuous learning is encouraged. Provide avenues for employees to report potential cybersecurity issues and ensure that there are open channels for them to seek clarification or further information.

In conclusion, designing and implementing a cybersecurity training program is about more than just imparting knowledge. It's about cultivating a culture of security awareness and vigilance, where every individual understands their role in safeguarding the organisation's digital assets. It is a continuous and collaborative effort that requires understanding, engagement, and, above all, commitment to a safer cyber environment.

9.5.3. Measuring the Effectiveness of Cybersecurity Training

Measuring the effectiveness of cybersecurity training is as essential as conducting the training itself. Yet, it's a complex process that involves several key facets. It is an exercise in both qualitative and quantitative assessment, ultimately guiding the iterative process of training refinement, based on the program's success in fostering better security habits and promoting an informed security-conscious culture.

The starting point often involves the use of pre- and post-training tests. By evaluating staff members' cybersecurity knowledge before and after the training program, one can quantitatively measure the increase in understanding. While test scores are not the ultimate indicator of cybersecurity behaviour, they do provide a helpful baseline to gauge the immediate impact of the training program.

Alongside knowledge, behaviour change is a crucial marker of a training program's effectiveness. Observing and tracking changes in employee behaviour can provide valuable insights. For instance, a reduction in the number of successful phishing attempts, or an increase in the number of reported suspicious emails can signal positive behavioural changes. Likewise, regular password updates or correct handling of sensitive data may reflect improved cybersecurity habits.

Surveys and feedback forms are another tool to assess the program's effectiveness. They can capture employees' perceptions of the training, its relevancy, clarity, and overall value. Anonymous surveys also offer an opportunity for staff to share their experiences, concerns, or suggestions, providing invaluable insights to refine the training approach.

It's also crucial to measure the training program's long-term impact. Cybersecurity isn't a 'set and forget' type of discipline. Regular follow-ups or refresher courses can help to determine if the learned behaviours are being maintained over time. Tracking key metrics over several months can provide a clearer picture of the training's sustained impact.

An essential, yet often overlooked aspect of effectiveness measurement is the impact on the organisation's risk profile. A successful training program should translate into a reduced risk of cyber incidents. Therefore, a decline in security incidents or breaches can be a powerful indicator of the program's effectiveness. Comparing incident frequency and severity before and after the implementation of the training program can offer a real-world measure of the program's success.

Finally, it is important to remember that cybersecurity is an ever-evolving field. The goalposts continually move as new threats and vulnerabilities emerge. It is vital to continually re-evaluate the effectiveness of the cybersecurity training program in light of these changes. Regular evaluations, updates, and adaptations ensure that the training program remains relevant, engaging, and above all, effective in equipping employees with the skills and knowledge necessary to navigate the dynamic cybersecurity landscape.

The essence of measuring the effectiveness of a cybersecurity training program lies not only in the metrics but also in a commitment to continuous improvement. It's about the process as much as the results. It's about learning, adapting, and always striving for a more secure, more informed, and more resilient organisation.

CHAPTER 10. LEGAL AND ETHICAL ASPECTS OF CYBERSECURITY

Chapter 10 of this cybersecurity guidebook provides a deep dive into the often complex, yet essential, legal and ethical aspects of cybersecurity. This realm of discussion, while laden with jargon and at times abstract, can have very tangible implications for individuals, businesses, and societies.

As cybersecurity becomes an increasingly crucial part of our digital world, legal and ethical considerations are moving to the forefront of discussions on how we manage and respond to cyber threats. These considerations form an intricate web that governs everything from individual privacy rights to international laws, and ethical standards in the industry. They represent the framework within which cybersecurity operates and the rules that it must abide by, forming an integral part of any comprehensive understanding of the field.

The legal aspect of cybersecurity encompasses the laws and regulations designed to protect people and companies from cyber threats and to prosecute those who perpetrate these crimes. Understanding the legal landscape is crucial not just for law enforcement agencies and legal professionals, but also for business leaders, IT professionals, and indeed anyone who operates in the digital space. Laws vary greatly around the world, and with cyberspace often transcending physical borders, the legal aspects of cybersecurity can be particularly complex.

Equally important are the ethical aspects of cybersecurity. These refer to the moral principles that guide our actions in this space. Ethics in cybersecurity is not merely about preventing hackers from stealing information or disrupting services. It's also about how businesses handle customer data, how governments balance national security with individual privacy, and how IT professionals make decisions that could impact people's digital lives.

In this chapter, we will navigate the nuanced labyrinth of legalities and ethics in the cybersecurity arena. We'll delve into topics like data privacy laws, cybercrime legislation, and international cybersecurity agreements. On the ethics front, we'll discuss the fine balance between security and privacy, the responsibilities of cybersecurity professionals, and the ethical considerations surrounding emerging technologies. This chapter will offer a comprehensive understanding of the importance of legal and ethical aspects in shaping the current and future state of cybersecurity.

While the topic may seem daunting, understanding the legal and ethical aspects of cybersecurity is not a choice but a necessity in today's digital era. This chapter aims to demystify these critical areas, offering clarity and insight, and equipping readers with the knowledge they need to navigate the cybersecurity landscape responsibly and ethically.

10.1. Overview of Cybersecurity Laws and Regulations

Section 10.1 of this chapter brings into focus the myriad of cybersecurity laws and regulations that exist around the globe. The purpose of these laws and regulations is twofold: on one hand, they aim to deter individuals and organisations from engaging in harmful cyber activities, and on the other hand, they aim to establish the legal framework for protecting and ensuring the cybersecurity of individuals, businesses, and nations.

In the ever-evolving landscape of the digital world, cybersecurity laws and regulations play a paramount role in delineating the rights, responsibilities, and protections associated with the use of cyberspace. While the specific laws and regulations may vary from one jurisdiction to another, they all share a common objective: to create a safer and more secure digital environment.

Cybersecurity laws and regulations cover a broad spectrum of issues. From rules regarding data privacy and protection, to provisions against cybercrimes such as hacking, identity theft, and cyberstalking. They also outline penalties for non-compliance, which can range from hefty fines to imprisonment, depending on the gravity of the breach and the jurisdiction in question.

In this section, we'll guide you through the intricate network of existing cybersecurity laws and regulations. Our aim is to shed light on their key provisions, their implications for individuals and businesses, and their significance in the global cybersecurity landscape. This knowledge will enable you to better understand your rights and responsibilities in cyberspace, as well as the potential legal ramifications of cyber activities.

It's essential to bear in mind that while this section offers an overview of the cybersecurity laws and regulations, it is not exhaustive. Given the dynamic nature of cyber threats and the digital world, these laws and regulations are continually being revised, updated, and supplemented. Additionally, new laws and regulations are constantly being proposed and enacted to address emerging issues in the cyber domain.

As we embark on this exploration of cybersecurity laws and regulations, keep in mind that understanding these laws isn't just a prerequisite for legal practitioners or policymakers. In our digital age, where most of our lives and businesses have a significant online footprint, this understanding is a fundamental necessity for everyone.

10.1.1. International Laws and Regulations on Cybersecurity

The international landscape of laws and regulations on cybersecurity is a mosaic, reflecting both the globally interconnected nature of cyberspace and the multitude of national jurisdictions it transcends. This intricate web of legislation plays a pivotal role in defining legal responsibilities and enforcing cybersecurity norms worldwide.

To comprehend the necessity and impact of international laws and regulations on cybersecurity, one must first acknowledge the borderless nature of cyberspace. The digital domain's ubiquitous nature necessitates a universal approach to manage security risks, creating a shared responsibility across nations. This situation has led to the emergence of various international laws, regulations, and guidelines designed to enhance cybersecurity globally.

One influential international standard in cybersecurity is the General Data Protection Regulation (GDPR) enforced by the European Union. Introduced in 2018, GDPR seeks to protect the personal data of EU residents and regulates its collection, processing, and movement. It has far-reaching implications, as it applies to any organisation, regardless of its geographical location, that deals with the data of EU citizens. Non-compliance can lead to hefty fines, making it a potent tool in ensuring data privacy and security.

Similarly, other regions have crafted their own regulations to protect the integrity, confidentiality, and accessibility of information within their jurisdictions. For instance, the United States employs multiple laws such as the Health Insurance Portability and Accountability Act (HIPAA) and the California Consumer Privacy Act (CCPA) that cater to different sectors and states respectively.

On the international level, the Budapest Convention is another milestone. It is the first international treaty aimed at combating cybercrime by harmonising national laws, improving investigative techniques, and increasing cooperation among nations. It also tackles the criminalisation of various cybercrimes such as illegal data interference, system interference, and infringement of copyright.

Despite these advancements, implementing international cybersecurity laws presents several challenges. The varying legal systems, cultural norms, and geopolitical interests across nations complicate the harmonisation of laws. Enforcement is another hurdle, as it often involves cross-border cooperation and extradition, both of which can be contentious issues.

Moreover, the rapidly evolving nature of cyber threats makes it difficult for the law to keep up. Legislators must balance the need to be flexible and forward-looking, to account for future technological advancements, while ensuring the legislation is specific enough to enforce and prevent misuse.

The establishment of robust international laws and regulations is a proactive response to cybersecurity threats. It promotes a safer, more secure digital world, where respect for privacy, integrity, and availability of information is not just an expectation but a legally binding obligation. As threats continue to evolve, so must the laws and regulations governing cyberspace. A shared commitment to cybersecurity and cooperation between nations is the need of the hour.

10.1.2. Role of Law Enforcement in Cybersecurity

In the ever-evolving landscape of cybersecurity, the role of law enforcement is critical. Law enforcement agencies serve on the front lines, mitigating the risks and threats, investigating incidents, and ensuring that the responsible parties are brought to justice. Their role spans across local, national, and international jurisdictions, presenting unique challenges and opportunities in the digital age.

Law enforcement agencies have a crucial role in responding to cyber incidents. When a breach occurs, they are among the first to be called upon to investigate. The primary goal here is to identify the perpetrators, gather evidence, and facilitate their prosecution. This responsibility requires sophisticated digital forensic skills to trace digital footprints, analyse logs, and recover data. Given the often-transnational nature of cybercrime, this investigative role often necessitates collaboration with law enforcement agencies from other jurisdictions.

In addition to reacting to incidents, law enforcement agencies also play a critical role in prevention and deterrence. They work closely with businesses, government agencies, and individuals to provide guidance on best practices in cybersecurity. This outreach helps enhance public awareness about the potential threats and the necessary precautions to take against them. Public awareness campaigns, training programs, and information sharing are some strategies that law enforcement uses to improve overall cybersecurity hygiene in the community.

Law enforcement's role extends into the realm of policy and legislation as well. They often contribute to the formation of laws and regulations pertaining to cybersecurity, providing insights into the realities of the digital battlefield. These contributions ensure that laws remain relevant and enforceable and help strike a balance between security needs and civil liberties.

Another area where law enforcement is increasingly active is in the development and application of cybersecurity technologies. Technologies such as encryption, intrusion detection systems, and blockchain are being adopted and modified by law enforcement to enhance their capabilities in tackling cybercrime.

However, the role of law enforcement in cybersecurity is not without challenges. The fast pace of technological change, the lack of sufficient trained personnel, and the jurisdictional boundaries in an inherently borderless cyberspace all complicate their work. Overcoming these challenges requires a global approach, with stronger international cooperation and investment in training and resources.

In essence, law enforcement agencies are a crucial pillar in the cybersecurity ecosystem. Their roles in investigation, prevention, policy formulation, and technological advancement are integral to maintaining the integrity and security of cyberspace. As the digital world continues to grow and evolve, so too will the roles and responsibilities of law enforcement in preserving its security and ensuring the rule of law prevails.

10.1.3. Legal Consequences of Cybersecurity Breaches

The legal consequences of cybersecurity breaches are substantial, far-reaching, and varied. They extend beyond immediate financial loss and can have severe implications for the reputation and long-term viability of an organisation. In an era where data is a vital asset, understanding these legal implications is crucial.

A cybersecurity breach could result in violation of various privacy laws and regulations, leading to hefty penalties. For instance, under the General Data Protection Regulation (GDPR) in the European Union, fines for non-compliance can go up to €20 million, or 4% of the firm's worldwide annual revenue of the prior financial year, whichever is higher. Likewise, violation of the Health Insurance Portability and Accountability Act (HIPAA) in the U.S. can result in fines up to $1.5 million per year for each violation. These laws mandate stringent standards for data protection and impose severe penalties for non-compliance.

Moreover, cybersecurity breaches often lead to costly legal actions, including lawsuits by affected customers, business partners, or employees. In the event of a data breach, organisations may face class-action lawsuits for failing to protect personal data adequately. In addition to direct damages, they may be liable for the cost of credit monitoring services for affected individuals and potential punitive damages.

Furthermore, organisations may face regulatory actions from governmental bodies. If a company is publicly traded, a significant cybersecurity incident needs to be disclosed to shareholders, which may affect the company's stock value. In some cases, executives may be held personally liable if they failed to take reasonable measures to prevent a breach or did not disclose it in a timely manner.

Additionally, cybersecurity breaches can also lead to contractual penalties. Many business contracts include clauses about data protection and privacy. A breach could be deemed as a failure to meet these contractual obligations, leading to termination of contracts and associated penalties.

Finally, beyond direct legal consequences, a cybersecurity breach can result in severe reputational damage. Trust is an invaluable asset, and once lost, it is tough to regain. In the aftermath of a breach, organisations may find their reputation tarnished, resulting in loss of business, a decrease in share value, and increased scrutiny from regulators and the public.

In conclusion, the legal implications of cybersecurity breaches are far-reaching, potentially resulting in financial penalties, lawsuits, regulatory actions, contractual penalties, and reputational damage. Given these potential legal consequences, organisations must invest in strong cybersecurity measures and incident response plans to mitigate the risk of breaches and handle them effectively when they occur.

10.2. Ethical Hacking and Its Role in Cybersecurity

In the high-stakes world of cybersecurity, where threats are multifarious and constantly evolving, it is crucial to stay one step ahead of potential attackers. This is where ethical hacking, also known as penetration testing or white-hat hacking, assumes a pivotal role. This practice is built on the premise that to outsmart a hacker, one must think like a hacker.

The advent of ethical hacking has considerably transformed the cybersecurity landscape, adding a proactive and anticipatory dimension to the traditionally defensive security approaches. Ethical hackers, equipped with the same technical skills as malicious hackers but bound by legal and moral constraints, are authorised to exploit vulnerabilities in an organisation's systems. Their purpose is not to create chaos or steal information, but to reveal weaknesses that can be fortified before a less benevolent hacker can exploit them.

Ethical hacking serves as an audit or evaluation of digital infrastructure, a rigorous test of an organisation's ability to withstand cyber-attacks. This proactive strategy bridges the gap between theoretical security implementations and their practical effectiveness. Without this evaluation, organisations are left to rely solely on their understanding of the systems and the hope that their defences will hold up against actual attacks.

However, the realm of ethical hacking is not without complexities. It requires a delicate balance to ensure that these skilled individuals use their abilities for the good of cybersecurity, and not its downfall. Therefore, ethical hacking is bound by stringent rules, codes of conduct, and legal agreements to ensure that all actions are conducted within the boundaries of legality and ethics.

In summary, ethical hacking is an indispensable element in today's cybersecurity strategies, underpinning the integrity and resilience of organisations in the face of growing digital threats. As we delve further into this chapter, we will explore the principles, practices, challenges, and benefits associated with ethical hacking in a comprehensive manner.

10.2.1. The Ethics of Ethical Hacking

Ethical hacking occupies a nuanced position in the cybersecurity ecosystem, straddling the thin boundary between permissible security enhancement activities and the realm of illegal intrusions. Ensuring that ethical hacking adheres to the 'ethical' part of its label requires a conscientious focus on the moral responsibilities that accompany the privileges of such cybersecurity role. This discussion will elucidate the ethical dimensions of ethical hacking and the principles that guide these professionals in their tasks.

At its core, ethical hacking is predicated on trust and responsibility. Organisations grant significant trust in ethical hackers, allowing them to probe the vulnerabilities of their systems. In return, ethical hackers bear the weighty responsibility to use their skills and access rights for the sole purpose of improving the organisation's security posture, and not for personal gain or unauthorised disclosure of sensitive information.

Ethical hackers also operate under the principle of legality. They must ensure their activities align with relevant local, national, and international laws, as well as organisational policies. This includes obtaining explicit, often written, permission from the relevant authority before initiating any testing or hacking activities and following defined rules of engagement.

Transparency is another cornerstone of ethical hacking. Ethical hackers must provide clear and detailed reports of their actions, findings, and suggestions for improvements. This transparency allows the organisation to fully understand the discovered vulnerabilities and the necessary remediation measures. Moreover, it also allows for accountability, a vital aspect in maintaining the trust relationship between ethical hackers and their employers.

The concept of 'do no harm' also applies to ethical hacking. Ethical hackers must strive to ensure that their actions do not result in damage to the systems they are testing or cause unnecessary disruption to business operations. Furthermore, they are also required to respect privacy and confidentiality, handling sensitive information with care and preventing unauthorised access or disclosure.

Despite the name, ethical hacking is not a carte blanche for cybersecurity professionals to do as they wish in the name of security. There are limitations and boundaries that should not be crossed, such as engaging in social engineering without the consent of the individual involved or hacking back against perceived attackers.

Finally, ethical hacking also includes a responsibility to contribute to the cybersecurity community. This involves sharing knowledge and insights, participating in professional development, and contributing to the development of best practices and standards. However, such contributions must be done in a manner that does not violate confidentiality agreements or disclose sensitive information.

In conclusion, the ethics of ethical hacking go beyond mere technical skill or the ability to find and exploit vulnerabilities. It involves a commitment to principles such as trust, responsibility, legality, transparency, and respect for privacy. Ensuring that ethical hackers adhere to these principles is crucial in maintaining the legitimacy and reputation of ethical hacking as a profession and as a vital component of cybersecurity defence.

10.2.2. The Process of Ethical Hacking: Reconnaissance, Scanning, Gaining Access, Maintaining Access, and Covering Tracks

Ethical hacking, also known as penetration testing or white-hat hacking, represents an essential element in the cybersecurity landscape, wielding a structured methodology to discover and patch vulnerabilities before they're exploited maliciously. Let's delve into the various stages of the ethical hacking process, namely: Reconnaissance, Scanning, Gaining Access, Maintaining Access, and Covering Tracks, and understand the implications of each in ensuring a resilient security posture.

The process begins with Reconnaissance, a pre-attack phase and the foundation of an ethical hacking operation. Ethical hackers, armed with a defined set of targets, gather as much information as possible about the systems they are to test. This can include identifying IP addresses, understanding the network topology, learning about the types of hardware and software in use, and discovering potential points of vulnerability. This step is often a blend of passive and active methods, from exploring publicly available data to more direct probing.

Scanning follows reconnaissance, where the ethical hacker uses automated tools to map out system networks, analyse open ports, understand system architecture, and inspect for any vulnerabilities. This step often employs software like Nmap, Nessus, or Wireshark to gain a better picture of the system's security posture. The results provide a roadmap for the next phase.

Gaining Access is where the ethical hacker attempts to exploit identified vulnerabilities to break into the system. This could be through techniques like SQL injection, cross-site scripting, or even social engineering methods. The aim is not to cause damage but to ascertain the feasibility of such an intrusion and understand the potential impact if a malicious actor were to exploit the same vulnerabilities.

Maintaining Access requires the ethical hacker to attempt to retain the access privileges they've gained, often by installing a backdoor in the system. This step is critical to understanding if an attacker can maintain long-term access to the system, posing a significant risk to the organisation's security posture.

Finally, Covering Tracks involves the ethical hacker attempting to remove any traces of the intrusion. This helps to ascertain if a malicious actor could exploit a vulnerability, maintain access, and then exit without being detected, thereby opening the system to repeated intrusions over time.

Each step of this process should be accompanied by detailed reporting, recording what actions were taken, what vulnerabilities were discovered, the severity of those vulnerabilities, and recommended mitigation strategies. These reports are crucial tools for organisations to understand their security posture, prioritise remediations, and implement strategic enhancements to their defences.

It's important to reiterate that all these steps must align with the ethical hacking principles: gaining explicit permission, doing no harm, and maintaining the confidentiality of any sensitive data encountered. This adherence to ethical boundaries is what fundamentally separates ethical hacking from black-hat hacking, turning potential cyber threats into cybersecurity's line of defence.

10.2.3. Building an Ethical Hacking Team: Skills, Tools, and Mindset

Assembling a competent ethical hacking team is not merely about technical proficiency. It involves a balanced mix of specialised skills, the correct toolkit, and, importantly, a mindset that's continually curious, security-focused, and firmly ethical. This team can play an integral role in an organisation's cybersecurity defence, proactively uncovering vulnerabilities and providing critical insights to strengthen security posture.

Skillset is the backbone of any ethical hacking team. Members should possess strong foundational knowledge of networking, programming, and system operations. They should understand common attack vectors and be familiar with the latest cybersecurity threats. Specific skills can range from penetration testing, reverse engineering, cryptography, to proficiency in various programming languages like Python, C++, or Java. Each member can specialise in a particular area, creating a diverse team that can tackle a wide array of security challenges.

In addition to individual skills, effective teamwork and communication are essential. The ability to articulate technical findings to non-technical stakeholders is invaluable. It ensures that the insights generated by the team's activities are properly understood and acted upon, enhancing the overall security awareness within the organisation.

The right tools can multiply the effectiveness of an ethical hacking team. There is an array of both open-source and commercial tools available that can aid in various stages of ethical hacking. Nmap, Wireshark, Metasploit, Burp Suite, and Nessus are some of the common tools used by ethical hackers for tasks like network mapping, vulnerability scanning, penetration testing, and traffic analysis.

Importantly, having the right mindset completes the ethical hacking team. This includes an inherent curiosity that drives them to continually learn and keep up with the evolving cybersecurity landscape. They should possess the determination to dig deep into systems and networks, leaving no stone unturned in the search for potential vulnerabilities.

Ethical integrity is paramount, given the sensitive nature of the information they are often privy to. It's this ethical underpinning that ensures their activities remain within the approved boundaries, maintaining the trust and confidence of stakeholders.

Lastly, a healthy dose of creativity can help the team think like potential attackers, enabling them to anticipate unconventional attack vectors and design tests that truly challenge the security mechanisms in place.

In essence, the blend of the right skills, tools, and mindset can empower an ethical hacking team to become an organisation's vanguard against cyber threats. It equips them to meet their overarching goal – to find and fix vulnerabilities before they can be maliciously exploited, thereby strengthening the organisation's cybersecurity resilience.

10.3. Privacy Considerations in Cybersecurity

As digital transactions continue to surge, vast amounts of personal data traverse global networks every second, creating an immense challenge for organisations. Ensuring privacy while maintaining robust cybersecurity measures is a critical balancing act that organisations must manage.

The preservation of privacy is a cornerstone of trust between businesses and their customers. When individuals entrust their personal data to an organisation, they implicitly trust that this information will be used responsibly, and not exposed, mishandled, or abused. However, given the attractiveness of personal data to cybercriminals, organisations are constantly at risk of data breaches.

In the context of cybersecurity, the idea of privacy extends beyond just protecting data from unauthorised access. It encompasses measures to ensure data confidentiality, integrity, and availability—often summed up as the CIA triad in cybersecurity. Confidentiality ensures that information is accessible only to authorised individuals. Integrity involves protecting data from unauthorised alteration, while availability guarantees that data is accessible to authorised users when needed.

Privacy-by-design, a principle where privacy safeguards are integrated into technology at the design stage, has gained traction in recent years. This involves incorporating privacy controls into the very architecture of systems and processes, rather than treating them as an afterthought. Such a proactive approach can help in preventing privacy breaches from happening in the first place.

Another aspect to consider is the minimisation of data collection. This principle advocates for collecting only the data that is necessary for a specific purpose, limiting the potential damage in the event of a breach. Additionally, deploying robust encryption mechanisms can further secure data in transit and at rest, making it harder for unauthorised users to access it.

In parallel, there is a growing emphasis on transparency about data practices. This involves being open with users about what data is being collected, how it is being used, and who it is being shared with. Such openness can help build trust and empower users to make informed decisions about their data.

It's also important for organisations to be aware of and comply with various data protection regulations, like the General Data Protection Regulation (GDPR) in Europe, the California Consumer Privacy Act (CCPA) in the U.S., and other similar laws worldwide. These regulations have stringent provisions for data privacy and impose heavy penalties for non-compliance.

The advent of new technologies like blockchain and homomorphic encryption also promise to enhance privacy protections by providing secure, decentralised ways to manage and store data. However, these technologies are still in their nascent stages and their wider adoption will require further research and standardisation.

10.3.1. The Intersection of Privacy and Cybersecurity

When we delve into the intersection of privacy and cybersecurity, we find an intricate matrix of responsibilities and challenges. Both fields aim to

354

protect information, but they do so from different angles, and their overlap creates a distinctive space where the two concepts meet and interact.

Privacy, in essence, is about controlling one's data. It's about having the right and the capability to decide who can access your personal information, under what conditions, and for what purposes. It's about maintaining a sphere of personal autonomy and preserving human dignity in an increasingly digitised world.

Cybersecurity, on the other hand, is the practice of protecting systems, networks, and programs from digital attacks. It's about erecting barriers against unauthorised access, ensuring data integrity, and establishing systems for recovery when breaches occur.

When these two domains intertwine, a new layer of complexity emerges. Cybersecurity can be seen as a means to an end for privacy. Without robust cybersecurity, privacy is at risk. If a system is compromised, personal data can be exposed, misused, or stolen, violating the privacy of individuals. Thus, proper cybersecurity practices are vital to privacy preservation.

However, the relationship between privacy and cybersecurity is not always straightforward. For instance, many cybersecurity measures involve monitoring network traffic or user behaviour to identify suspicious activities. Such monitoring can potentially infringe on privacy if not carefully managed. Hence, a balance needs to be struck between security needs and privacy rights.

Organisations need to employ strategies like anonymisation and pseudonymisation to protect user privacy while still enabling security monitoring. Anonymisation refers to the process of removing personally identifiable information from data sets, making it impossible to identify individuals, while pseudonymisation replaces identifying fields within a data record with artificial identifiers or pseudonyms.

Further, the adoption of a privacy-by-design approach is essential. This means incorporating privacy safeguards right from the design stage of any product or process. By making privacy an integral part of design, organisations can ensure that they do not have to choose between privacy and security, but instead, achieve both.

Privacy Impact Assessments (PIAs) are another tool at the intersection of privacy and cybersecurity. PIAs help organisations assess how a particular project or system may affect the privacy of individuals, enabling them to address potential issues before they occur.

In the realm where privacy and cybersecurity meet, legal compliance also plays a critical role. Numerous privacy laws worldwide, such as the GDPR in Europe and the CCPA in California, require organisations to implement adequate cybersecurity measures to protect personal data. Non-compliance can result in hefty penalties, making it a priority for organisations to ensure their cybersecurity measures support their privacy obligations.

In the final analysis, the intersection of privacy and cybersecurity is a dynamic space where legal, ethical, and technical aspects overlap. It requires a thoughtful and proactive approach to harmonise the demands of both privacy and security, creating a digitally secure environment that respects and upholds the individual's right to data privacy.

10.3.2. Privacy Laws and Regulations: GDPR, CCPA, and Beyond

The privacy landscape is not a static entity; it is continuously evolving, heavily influenced by the advent of new technologies, changing societal attitudes, and new legal frameworks. Various laws and regulations have been enacted worldwide to address growing concerns about personal data protection, among which the General Data Protection Regulation (GDPR) and California Consumer Privacy Act (CCPA) stand out as significant legislative pieces. They showcase how regions can shape the privacy landscape and provide a benchmark for other jurisdictions to follow.

The General Data Protection Regulation (GDPR) came into effect in the European Union (EU) in May 2018. Its broad scope, rigorous requirements, and potential for high financial penalties transformed the global privacy landscape almost overnight. The GDPR applies to any entity, anywhere in the world, that processes the personal data of individuals within the EU. It grants individuals numerous rights, including the right to access their data, correct inaccuracies, object to processing, and request data deletion under the right to be forgotten.

The GDPR also imposes stringent obligations on data controllers and processors. It requires them to implement appropriate technical and organisational measures to ensure data protection, adopt a privacy-by-design approach, conduct Data Protection Impact Assessments (DPIAs) for high-risk processing activities, and appoint a Data Protection Officer (DPO) under certain circumstances. The principle of accountability, which requires organisations to demonstrate their compliance with GDPR, is a cornerstone of the regulation.

On the other side of the Atlantic, the California Consumer Privacy Act (CCPA) represents a landmark in US privacy law. Effective from January 2020, the CCPA gives California residents various rights regarding their personal information, similar to the GDPR. These include the right to know what personal information is collected, used, shared, or sold, the right to delete personal information held by businesses, and the right to opt-out of the sale of personal information.

While the GDPR and CCPA have grabbed headlines, they are not the only significant privacy laws in effect. Brazil's General Data Protection Law (LGPD), India's Personal Data Protection Bill (PDPB), and China's Personal Information Protection Law (PIPL) represent major strides in privacy legislation in these countries. Each of these laws has its nuances, reflecting local cultural, political, and societal contexts, yet they all share a common goal of protecting personal data in an increasingly digital and interconnected world.

For organisations operating internationally, understanding and complying with these different laws can be challenging. A comprehensive privacy program that includes regular privacy assessments, data mapping, and ongoing staff training can help organisations navigate this complex regulatory landscape. Additionally, organisations should be prepared to adapt as new laws emerge and existing laws evolve. In this ever-changing terrain of privacy laws and regulations, continuous monitoring, adaptation, and a commitment to privacy excellence are key to maintaining compliance and protecting individual privacy.

10.3.3. Privacy-by-Design in Cybersecurity

Privacy-by-design (PbD) is an approach to cybersecurity that embeds privacy considerations into every facet of a product or system during its entire lifecycle, from the initial design phase through to final decommission. It represents a proactive rather than reactive approach to privacy, incorporating privacy safeguards at the outset, instead of retroactively patching privacy vulnerabilities.

The core principle behind PbD is the notion that privacy is a fundamental right that must be intrinsically embedded within technologies, business practices, and networked infrastructures. It underscores the importance of building in privacy up front, ensuring that default settings of a system or service are automatically privacy protective. This way, privacy becomes an integral, built-in component of systems and practices, rather than an add-on feature.

PbD covers several fundamental principles. It encourages 'end-to-end security,' meaning that information should be secured across its entire lifecycle from collection through to disposal. It promotes visibility and transparency to assure stakeholders that operations are legitimate and that their privacy is being protected. It also prioritises respect for user privacy by keeping the interest of individuals at the forefront.

Implementing PbD in cybersecurity involves several key steps. First, privacy objectives and requirements should be defined at the earliest stages of product or system design. These requirements should align with applicable laws, regulations, and standards. Second, privacy impact assessments (PIAs) can be performed to identify potential privacy risks and develop mitigation strategies. Third, privacy controls should be integrated into the technical design and architecture of systems. Finally, privacy policies and procedures should be continuously monitored, reviewed, and updated to respond to changes in the environment and threats.

Applying a PbD approach to cybersecurity not only helps organisations comply with privacy laws and regulations, but it also builds trust with customers and stakeholders. As privacy concerns continue to grow in a data-driven world, privacy-by-design is becoming an increasingly important principle in cybersecurity. Integrating privacy into the design and operation of systems and practices can significantly reduce privacy risks and foster a culture of privacy awareness within an organisation. This proactive approach can help avoid costly and damaging privacy breaches while protecting one of the most important rights of individuals: the right to privacy.

10.4. Dealing with Cybersecurity Breaches

Despite the most robust preventative measures, the rapidly evolving and increasingly sophisticated threat landscape necessitates a focus not just on prevention, but also on preparedness for inevitable incidents. Addressing cybersecurity breaches involves a strategic balance between proactive prevention, swift and effective response, and a continuous learning approach to minimise future incidents.

Cybersecurity breaches can disrupt operations, damage reputations, and lead to significant financial loss. They can also potentially have severe legal and regulatory consequences, with data protection legislation globally imposing stringent penalties for breaches. It is vital, then, that every organisation, irrespective of size or sector, develops a comprehensive incident response plan to manage and mitigate the impact of cybersecurity breaches.

In the heat of a breach, an organisation's ability to react effectively is often determined by the level of planning and preparation beforehand. An incident response plan typically outlines how to identify, contain, eradicate, and recover from a breach, as well as the steps to conduct a thorough post-incident analysis. This plan must be clear and actionable, laying out responsibilities, communication strategies, and specific remediation steps.

The incident response team, often composed of professionals from diverse backgrounds such as IT, legal, public relations, and human resources, is instrumental in executing this plan. Regular training and simulated cyber-attack exercises can help this team respond effectively under the pressure of an actual breach scenario.

One of the critical tasks when a breach occurs is to identify the compromised systems quickly and prevent further intrusion. This containment phase may involve isolating affected networks, changing passwords, or even temporarily shutting down services. Following containment, the eradication phase involves eliminating the threat, while the recovery phase involves restoring services, data, and operations to their pre-breach states.

Transparency during a breach is also vital. This includes clear and timely communication with stakeholders, from employees and customers to regulatory bodies. Regulatory compliance is essential, as laws such as GDPR mandate reporting breaches within a certain timeframe.

Post-breach, it's crucial to conduct a thorough analysis to understand the breach's cause, its impacts, and how it was handled. This process provides valuable lessons to prevent similar occurrences in the future and improve the incident response plan.

Cybersecurity breaches are an unfortunate reality in today's digital age. However, through effective preparation, swift action, and a commitment to continuous improvement, organisations can significantly reduce their impact and turn a potential disaster into a manageable incident.

10.4.1. Legal Obligations After a Cybersecurity Breach

Once a cybersecurity breach occurs, the path to recovery is often complicated and fraught with numerous legal obligations. These obligations differ based on the jurisdiction in which an organisation operates, the sector in which it functions, and the nature of data it handles. Meeting these legal obligations is crucial not just to avoid penalties, but also to manage reputation risks and maintain customer trust.

In the immediate aftermath of a breach, one of the primary legal obligations is breach notification. Many jurisdictions have regulations requiring companies to report data breaches to both the affected individuals and regulatory authorities. The General Data Protection Regulation (GDPR) in Europe, for instance, mandates companies to report breaches to the appropriate supervisory authority within 72 hours of discovery. The notification must include details about the nature of the breach, the categories and approximate number of individuals affected, the potential consequences, and the measures taken or proposed to mitigate its effects.

Furthermore, in the United States, most states have laws requiring businesses to notify affected individuals about breaches involving personally identifiable information. Similarly, in Australia, the Notifiable Data Breaches (NDB) scheme under the Privacy Act necessitates businesses to notify individuals affected by a data breach that could potentially result in serious harm.

Additionally, there may also be sector-specific reporting requirements. For instance, the Health Insurance Portability and Accountability Act (HIPAA) in the U.S requires healthcare entities to report breaches affecting 500 or more individuals to the Department of Health and Human Services, the affected individuals, and the media.

Beyond breach notification, organisations have a legal obligation to mitigate the impact of the breach. This typically involves identifying and addressing vulnerabilities that led to the breach and taking steps to prevent future incidents. If a company is found to have insufficient security measures in place, it could face penalties and legal action from both regulators and affected individuals.

Organisations are also required to cooperate with law enforcement and regulatory investigations into the breach. This can involve providing access to systems, logs, and other relevant data, as well as making personnel available for interviews.

Organisations must remember that these legal obligations are not just about compliance – they are about demonstrating accountability, protecting stakeholders, and maintaining trust in the digital economy. The aftermath of a cybersecurity breach can indeed be challenging, but meeting these legal obligations can help organisations navigate the path to recovery effectively.

10.4.2. Communicating Cybersecurity Breaches to Stakeholders

In a highly connected business environment, a cybersecurity breach can significantly impact various stakeholders, including customers, employees, shareholders, suppliers, and regulators. Communicating with these stakeholders in a timely, transparent, and empathetic manner is essential to manage the crisis effectively, maintain trust, and demonstrate accountability.

The process of communicating a cybersecurity breach to stakeholders begins with a clear understanding of what happened, including the nature and scope of the breach, the data or systems affected, and the steps being taken to address the situation. This understanding helps create a coherent and consistent message that accurately informs stakeholders without causing unnecessary panic.

The next step involves identifying the specific messages for each group of stakeholders. For customers and employees, for example, the communication may focus on what information was compromised, how it might affect them, and what they can do to protect themselves, such as changing passwords or monitoring their financial accounts. It may also include what the organisation is doing to prevent a recurrence, such as enhancing its security measures or working with law enforcement.

Communicating with shareholders and investors may require more detail about the financial impact of the breach, including any potential liabilities, costs associated with remediation, and how the incident may affect business operations and the company's overall financial position. It's also important to address the steps being taken to manage the crisis, protect the company's assets, and safeguard its reputation.

Suppliers and business partners also need to be informed, especially if the breach could affect them. This can help maintain business continuity, ensure continued collaboration, and demonstrate the company's commitment to transparency and accountability.

Finally, regulators must be informed as per legal requirements. The communication should include the details of the breach, its impact, and the steps taken to address it.

The channels used for communicating a cybersecurity breach can vary based on the stakeholder group. They may include emails, letters, phone calls, press releases, social media posts, or a combination thereof. Whichever channels are chosen, it's crucial to ensure that the communication is clear, concise, and consistent.

Importantly, communicating a cybersecurity breach is not a one-time activity. It involves continuous updates as new information emerges and as the company makes progress in addressing the breach. By effectively communicating with stakeholders during such a crisis, organisations can demonstrate their commitment to transparency, build trust, and help ensure a stronger recovery.

10.4.3. Post-Breach Investigation and Remediation

The aftermath of a cybersecurity breach is a critical time for any organisation. This period involves a complex process of investigating the cause and scope of the incident, identifying and rectifying vulnerabilities, and taking measures to prevent future occurrences. The post-breach investigation and remediation phase is the organisation's opportunity to learn from the breach and enhance its cybersecurity posture.

Investigating a breach begins with determining its cause and extent. Understanding how the breach occurred and what information was compromised is crucial to preventing similar incidents in the future. This might involve working with internal teams or bringing in external cybersecurity experts to dissect the attack. Forensic techniques can be used to analyse system logs, patterns of network traffic, and any suspicious or unusual activity leading up to the breach.

The investigation also includes identifying any vulnerabilities that were exploited during the attack. This could be weaknesses in software, hardware, or human behaviour, such as poor password practices or failure to follow security protocols. Understanding these vulnerabilities helps in taking targeted remedial actions.

Once the breach's cause and scope are understood, the next step is remediation. This involves taking immediate measures to secure the compromised systems and data. Actions could include patching software, changing access controls, strengthening network defences, and improving training for employees.

Moreover, as part of the remediation process, organisations also need to consider the legal and regulatory implications of the breach. This could involve reporting the breach to regulatory authorities, notifying affected customers, and complying with any legal obligations that arise from the breach, such as potential liabilities.

Once immediate risks are mitigated, organisations should focus on longer-term measures to enhance their cybersecurity posture. This could involve investing in advanced cybersecurity technologies, improving staff training, or enhancing the organisation's incident response plan based on lessons learned from the breach.

The post-breach phase also provides an opportunity to review and update the organisation's cybersecurity policy, aligning it with the current threat landscape and the organisation's risk tolerance. This process can help create a stronger, more resilient cybersecurity framework that can better withstand future threats.

In conclusion, a post-breach investigation and remediation is a critical process that helps organisations recover from a cybersecurity incident, learn from it, and strengthen their security controls. With the ever-evolving cyber threat landscape, having a robust and adaptive post-breach strategy is integral to managing and mitigating cybersecurity risks.

CHAPTER 11. FUTURE OF CYBERSECURITY

As we continue to digitise our economies, societies, and personal lives, our vulnerabilities in cyberspace expand in parallel. This makes the future of cybersecurity a complex yet fascinating topic, as it grapples with novel threats, evolving technologies, and ever-changing regulations. As we journey into this final chapter, we venture into the world of future possibilities, exploring the trends that will shape cybersecurity and how we can prepare for them.

The rapid development and deployment of technologies such as artificial intelligence (AI), machine learning (ML), Internet of Things (IoT), and quantum computing are reshaping the cybersecurity landscape. They hold enormous potential for improving security defences, but they also create new attack surfaces and vulnerabilities. Therefore, understanding how these technologies impact cybersecurity is crucial for future preparedness.

Moreover, the increasing prevalence of cyber warfare and state-sponsored cyber-attacks heighten the need for resilient national cybersecurity strategies. Cyber threats are no longer limited to the realm of individual organisations or consumers; they have broad geopolitical implications, necessitating a collective and coordinated global response.

On another note, cybersecurity's future will be shaped by regulatory evolution. With privacy concerns and data breaches on the rise, we can expect more stringent data protection laws and regulations, similar to the GDPR, in different parts of the world. Organisations will need to be proactive in their compliance efforts and understand the global regulatory landscape to avoid significant penalties.

Furthermore, cybersecurity education and workforce development will play a critical role in the future. As cyber threats become more sophisticated, there is a growing need for skilled cybersecurity professionals who can keep pace with the evolving threat landscape. Thus, initiatives to promote cybersecurity education, training, and career development will be integral to building a secure digital future.

Finally, organisations of the future will need to adopt a security-centric culture, where cybersecurity is not just an IT concern but a priority at all levels, from the boardroom to the front lines. This shift in mindset will be crucial to effectively manage cybersecurity risks in an increasingly interconnected and digital world.

As we delve into the depths of these topics in this chapter, we hope to equip you with a solid understanding of the future trends and challenges in cybersecurity. More importantly, we aim to provide insights that will enable you to navigate this rapidly changing landscape, safeguarding your data, systems, and ultimately, your future in the digital age.

11.1. Emerging Threats and Challenges

As we continue our voyage into the future of cybersecurity, understanding these emerging threats and the associated challenges becomes a crucial compass. Not only do these insights help organisations prepare for what lies ahead, but they also underscore the importance of continuous evolution in cybersecurity strategies and technologies.

The rise of artificial intelligence (AI) and machine learning (ML) has been a double-edged sword. On the one hand, they have significantly improved our ability to detect and mitigate threats. On the other, they have given rise to more sophisticated cyber-attacks. Adversarial AI, for instance, represents a class of threats where AI systems are used to outsmart security measures, highlighting the increasing complexity of future cyber-attacks.

Additionally, the Internet of Things (IoT), while offering unparalleled connectivity and convenience, has vastly expanded the attack surface for cybercriminals. IoT devices, often lacking robust built-in security, are increasingly targeted, leading to threats like botnet attacks. Future homes, cities, and industries teeming with IoT devices will necessitate innovative strategies to ensure security without compromising the benefits of connectivity.

Another emerging challenge lies in the realm of quantum computing. Quantum computers, when realised, could disrupt existing cryptographic systems, posing significant challenges to data security. Preparing for a post-quantum cryptography era is therefore a major task facing the cybersecurity community.

Furthermore, escalating geopolitical tensions have seen a rise in state-sponsored cyber-attacks. These cyber threats, often highly sophisticated, target not just government institutions but also private organisations and critical infrastructure. In the face of such advanced persistent threats (APTs), the need for strong national and international cybersecurity defences has never been greater.

Moreover, the rapid digitisation seen in the wake of the global pandemic has escalated the cyber risks associated with remote work. Organisations are grappling with how to secure their networks and data in an environment where traditional office perimeters no longer apply. The trend of 'work from anywhere' means that organisations must rethink their cybersecurity strategies, focusing on securing data and identities rather than just network perimeters.

Finally, the ongoing data privacy movement poses new challenges. Balancing the need for data-driven insights with respect for individual privacy rights is a delicate act that organisations must master. With the rise of privacy-focused regulations, failure to handle personal data properly can lead not just to reputational damage, but also hefty penalties.

In summary, the landscape of emerging threats and challenges in cybersecurity is as vast as it is diverse. However, by understanding these threats and taking proactive steps to mitigate them, we can harness the power of technology while minimising risks. The key is to remain vigilant, adaptable, and committed to maintaining robust cybersecurity practices as we venture into the future.

11.1.1. The Changing Cyber Threat Landscape: Predicted Trends

The cyber threat landscape has always been fluid, with the nature of threats continuously evolving and escalating. In recent years, we've seen cybercriminals shift from mass, opportunistic attacks to more targeted tactics, exploiting the vulnerabilities in emerging technologies and adjusting to societal changes. As we gaze into the future, understanding the predicted trends in this shifting landscape is essential to safeguard our digital lives and assets.

Firstly, the rapid development and adoption of artificial intelligence (AI) and machine learning (ML) in various industries, including cybersecurity, isn't going unnoticed by cybercriminals. It's predicted that these technologies will be weaponised to create AI-driven cyber-attacks, where malicious AI systems could learn from and bypass security defences, creating a high-stakes cat-and-mouse game between security teams and attackers. The cybersecurity industry must stay abreast of AI advances, researching and implementing robust countermeasures.

Secondly, the advent of quantum computing is a looming trend that may drastically reshape the threat landscape. The powerful computational capabilities of quantum computers could potentially crack existing cryptographic systems, undermining the confidentiality and integrity of digital communication and data storage. As quantum computing continues to mature, organisations need to invest in quantum-resistant cryptographic systems, a research field that is burgeoning but yet to be standardised.

Thirdly, state-sponsored cyber-attacks are becoming increasingly common and sophisticated, a trend expected to escalate in the future. These advanced persistent threats (APTs) have larger budgets, more resources, and are typically more adept at breaching defences and staying undetected. This reinforces the need for a more cooperative global approach to cybersecurity, with information sharing and unified defence strategies.

Fourthly, the accelerated shift towards remote work and increased reliance on cloud-based services due to the COVID-19 pandemic has expanded the attack surface for cybercriminals. These changes have resulted in a surge of attacks aimed at cloud infrastructure and services, and end-users who may not follow best security practices while working remotely. Organisations need to build more flexible and robust security strategies, focusing on aspects like zero-trust architecture and secure access service edge (SASE) model.

Fifthly, the explosive growth of Internet of Things (IoT) devices is likely to continue, creating new attack vectors. Given their notorious lack of built-in security, IoT devices present an attractive target for cybercriminals. Secure design and update mechanisms for IoT devices must be a priority for manufacturers.

Next, social engineering attacks, particularly phishing, remain a significant threat and are expected to evolve in their sophistication. The use of AI could make these attacks more effective, crafting highly personalised messages based on victims' online behaviour and preferences. Raising user awareness remains a key element of mitigating this risk.

Finally, the growing demand for privacy is leading to a rise in privacy-centric attacks, where cybercriminals could exploit privacy-preserving technologies to hide malicious activities. While organisations strive to enhance privacy measures, they must also ensure they don't inadvertently create blind spots in their monitoring capabilities.

In summary, the future of the cyber threat landscape is marked by an escalation in the scale and sophistication of attacks. Responding to these predicted trends requires proactive measures, constant learning, and adapting. The aim should be to build resilient systems that can not only withstand attacks but also anticipate them and adapt to protect the most valuable digital assets. Despite the challenges, a thoughtful approach to cybersecurity can turn these potential threats into opportunities for strengthening security posture.

11.1.2. Cybersecurity Challenges in Emerging Technologies

As we embrace emerging technologies, from artificial intelligence (AI) and quantum computing to the Internet of Things (IoT) and 5G connectivity, we also need to navigate the cybersecurity challenges these technologies bring along. Given the rapidly evolving nature of these technologies, their associated security issues are continually shifting, but a few key challenges can be anticipated.

Artificial Intelligence and Machine Learning (AI/ML) is a double-edged sword in cybersecurity. While AI/ML can dramatically improve threat detection and response times, the technology can also be used maliciously. Attackers are starting to use AI/ML to automate attacks, evade detection, and tailor social engineering attacks. This calls for continuous research to stay ahead in the AI arms race, employing AI/ML to counter these sophisticated threats while understanding and mitigating the risks posed by adversarial AI/ML.

Quantum computing, once it becomes a reality, poses a significant threat to existing cryptographic systems. The sheer computational power of quantum machines can theoretically break the encryption algorithms that protect sensitive data today. The cybersecurity industry has to focus on the development and standardisation of quantum-resistant cryptographic algorithms, an area that is currently at a nascent stage.

The Internet of Things (IoT) is proliferating at an incredible pace, expanding the attack surface exponentially. IoT devices, often designed with little or no security, can serve as easy points of entry for cybercriminals. The challenge lies in creating universal security standards for these devices, enhancing their built-in security, and ensuring secure update mechanisms to patch vulnerabilities promptly.

Similarly, the adoption of 5G technology brings new cybersecurity challenges. While 5G promises faster speeds and lower latency, it also introduces a more complex, software-driven network infrastructure that can be exploited. The network slicing feature of 5G, for instance, can be a potential avenue for attacks if not correctly secured. As we move towards this new standard, rigorous security testing and securing the software-defined networking components are crucial.

Blockchain technology, while often associated with security and trust, also poses certain challenges. The immutability characteristic of blockchain can make it a potential carrier for malicious content, and its decentralised nature can create regulatory challenges. The security of smart contracts, an essential feature in many blockchains, is another concern due to their potential vulnerabilities.

Cloud technologies, with the shift towards multi-cloud and hybrid cloud environments, also present a complex cybersecurity landscape. Maintaining visibility and control over data and services spread across multiple providers, and managing the security configurations of these services, are daunting tasks. A focus on security strategies specific to the cloud, like the secure access service edge (SASE) model, is needed.

Finally, the increasing adoption of Augmented Reality (AR) and Virtual Reality (VR) technologies opens up a whole new world of cyber threats. Cyberattacks in these spaces can potentially lead to physical harm, not just digital. Ensuring the security of these immersive technologies is still a largely uncharted territory.

In conclusion, while emerging technologies hold great promise, they also present new frontiers for cyber threats. Overcoming these challenges requires a proactive, adaptive approach to cybersecurity that evolves alongside these technologies. The interplay between technology advancement and security will be a defining aspect of the digital future.

11.1.3. The Implication of Cyber Warfare and State-sponsored Attacks

The dark underbelly of technological progress is the escalation of cyber threats at a global scale, increasingly manifesting as cyber warfare and state-sponsored attacks. This evolution in cybercrime signals a seismic shift in the security landscape and has profound implications on national security, economic stability, and privacy.

State-sponsored cyberattacks represent a new form of warfare in the digital age. Unlike traditional warfare, these attacks are not bound by geographical constraints or conventional battle lines. They are often highly sophisticated, well-resourced, and able to inflict significant damage on targeted nations' critical infrastructures such as energy grids, telecommunications, and financial systems.

One of the profound implications of such attacks is the potential for economic disruption. By targeting critical infrastructure or financial institutions, state-sponsored attackers can inflict massive economic damage. Additionally, these attacks can lead to the theft of valuable intellectual property, which can have long-term economic consequences and undermine competitiveness.

In the realm of national security, state-sponsored attacks present a clear and present danger. They can be leveraged to disrupt military operations, interfere with intelligence activities, and weaken a nation's defence capabilities. The increasing digitisation of military assets and communications makes this an area of paramount concern.

The advent of state-sponsored cyberattacks also raises significant privacy issues. Such attacks often involve widespread surveillance activities and the harvesting of vast amounts of personal data. This not only undermines individual privacy rights but can also be used to manipulate public opinion, as evidenced by the rise in influence operations.

Another troubling implication of state-sponsored attacks is the potential for escalation and the absence of clear norms and regulations. The line between espionage (generally considered acceptable state behaviour) and cyber warfare (which is not) is blurred in the digital realm. This lack of clarity, combined with the difficulty in attribution, creates an environment conducive to escalation.

In the face of these challenges, nations are compelled to bolster their cybersecurity postures. This includes strengthening the security of critical infrastructure, investing in advanced threat detection and response capabilities, and improving incident response.

Additionally, there is an urgent need for international cooperation in establishing norms of state behaviour in cyberspace, sharing threat intelligence, and facilitating coordinated responses to cyber threats. This could materialise in the form of bilateral agreements, multilateral treaties, or international regulatory bodies.

In conclusion, the increasing prevalence of cyber warfare and state-sponsored attacks is changing the face of global security. This necessitates a collective, collaborative, and comprehensive approach that encompasses technological innovation, policy development, legal reform, and international cooperation. Cybersecurity is no longer just an IT issue; it's a matter of international security and diplomacy.

11.2. The Role of AI and Machine Learning in Cybersecurity

As we traverse deeper into the digital age, artificial intelligence (AI) and machine learning (ML) are becoming increasingly integral to the future of cybersecurity. These advanced technologies, with their self-learning and predictive capabilities, offer remarkable opportunities for identifying and countering cyber threats, but they also introduce new vulnerabilities that can be exploited by cybercriminals.

AI and ML, due to their inherent ability to analyse vast quantities of data and recognise patterns, are revolutionising threat detection and response. Traditionally, cybersecurity teams have to manually comb through logs and alerts to identify potential threats, a task that is both time-consuming and prone to human error. With the advent of AI and ML, this process can be automated, and threats can be identified in real time with a high degree of accuracy. This fundamentally enhances the speed and efficiency of threat detection, allowing for immediate mitigation actions.

AI and ML can also play a pivotal role in predictive cybersecurity. By analysing historical data on cyber threats and attacks, these technologies can predict future attack patterns and trends. This allows cybersecurity teams to proactively develop defences and strategies, rather than merely reacting to threats as they occur. The predictive capabilities of AI and ML, therefore, significantly boost a company's resilience against cyber-attacks.

In addition to threat detection and prediction, AI and ML are increasingly being leveraged for incident response. Automated response tools powered by these technologies can take immediate action when a threat is detected, such as isolating affected systems or blocking malicious IP addresses. This considerably reduces the window of opportunity for attackers to inflict damage.

However, the integration of AI and ML in cybersecurity is not without challenges. As these technologies become more sophisticated, there is a growing risk of them being exploited by cybercriminals. For instance, attackers can use AI to automate their attacks, making them faster and more effective. Additionally, they can use ML to study and mimic the behaviour of legitimate users, making their activities harder to detect.

Moreover, the 'black box' nature of some AI and ML systems — where the decision-making processes are not easily understandable by humans — can pose accountability and transparency issues. This can create difficulties in complying with regulations that require transparency in data processing activities.

In conclusion, AI and ML are set to be game changers in cybersecurity, offering enhanced capabilities for threat detection, prediction, and response. However, their use also necessitates a careful consideration of the potential risks and ethical implications. As such, the future of cybersecurity will involve striking a balance between leveraging these technologies to bolster defences and ensuring they are used responsibly and ethically.

11.2.1. AI and ML in Threat Detection and Response

As the digital universe expands, generating enormous quantities of data each day, the task of manually reviewing every security log and alert for potential threats becomes an impractical endeavour. This is where artificial intelligence (AI) and machine learning (ML) enter the cybersecurity arena, revolutionising the methods and efficiency of threat detection and response.

AI and ML are capable of handling and analysing vast data troves in real time, a feat unattainable by human analysts. By learning from patterns and behaviours, these technologies are able to discern deviations indicative of potential cyber threats. This automation expedites threat detection, reduces false positives, and makes threat response more accurate, ultimately bolstering the defences of an organisation's cybersecurity infrastructure.

An important application of AI and ML in cybersecurity is anomaly detection. Anomalies, or deviations from established patterns, are often a tell-tale sign of malicious activity. Machine learning algorithms can be trained to learn normal user or system behaviour, and subsequently flag any deviations from this norm. This sort of intelligent and adaptive threat detection provides a significant advantage in the perpetual game of cat-and-mouse between cybersecurity professionals and cyber adversaries.

Moreover, AI and ML can contribute significantly to automated incident response, an essential component of modern cybersecurity frameworks. Upon detection of a cyber threat, AI-enabled systems can take immediate action without human intervention, including blocking IP addresses, isolating affected systems, and notifying incident response teams. This instantaneous response drastically reduces the 'dwell time' of threats, thereby limiting the potential damage.

On the flip side, as cybersecurity measures become smarter, so do the attack tactics deployed by cybercriminals. A growing concern is the potential for malicious actors to harness AI and ML to enhance their own arsenal. For example, AI could be used to conduct more effective phishing campaigns, by analysing behavioural data and crafting highly personalised and convincing deceptive emails. Cybersecurity teams must therefore remain vigilant and continue to innovate, in order to stay ahead of increasingly sophisticated threats.

A future-oriented view of cybersecurity must acknowledge and address these possibilities. The next generation of cybersecurity tools and strategies should not only focus on leveraging AI and ML for threat detection and response but also consider how to safeguard these systems from manipulation. An intricate understanding of how these technologies function, and where their vulnerabilities lie, will be vital in securing the cyber landscape of the future.

11.2.2. Ethical and Practical Considerations in AI-based Cybersecurity

The integration of AI and ML into cybersecurity is not just a technical endeavour but also an ethical and practical one. The implications are extensive, requiring thoughtful navigation to avoid pitfalls and harness the full potential of these technologies.

Starting with the ethical aspect, the use of AI and ML in cybersecurity invariably involves the handling of massive volumes of data, some of which could be highly sensitive or personal. As such, organisations must balance the imperative for robust cybersecurity with respect for individual privacy. The deployment of AI and ML should align with privacy regulations like GDPR and ethical guidelines to ensure data is handled responsibly and securely. For instance, principles such as data minimisation, purpose limitation, and transparency must be adhered to when deploying these technologies.

Moreover, AI-based cybersecurity systems can inadvertently introduce bias and lead to unfair outcomes. For example, a biased ML model might flag activities of certain user groups as suspicious based on flawed or discriminatory learning data. Hence, organisations must incorporate fairness, accountability, transparency, and ethics (FATE) principles when building their AI and ML models to ensure they don't inadvertently marginalise certain user groups.

From a practical perspective, the implementation of AI and ML is often complex and resource intensive. It requires a combination of domain expertise, data science skills, and computational resources. The black-box nature of some ML models also poses a challenge for cybersecurity teams. Interpretability is vital in cybersecurity, where understanding why a system flags a certain behaviour as anomalous could be as important as the detection itself. Consequently, the use of explainable AI (XAI) techniques can help in understanding the decision-making process of AI models and instil trust in their predictions.

AI and ML models also need to be robust against adversarial attacks. Sophisticated attackers might attempt to poison the training data or use other techniques to mislead the AI systems. Therefore, security must be embedded not just in the deployment, but also in the design and training phases of AI and ML models.

Lastly, there's the problem of overreliance on AI, which can lead to complacency among human analysts. It's crucial to remember that AI is a tool that aids cybersecurity professionals, not a substitute for their expertise. A blend of human intelligence and AI capabilities can create a more robust and adaptive cybersecurity framework. In conclusion, the successful integration of AI and ML in cybersecurity will depend on careful ethical and practical considerations, not just technological prowess.

11.2.3. Future Trends: Autonomous Cyber Defence Systems

As we stare into the future of cybersecurity, one concept seems to gather momentum: Autonomous Cyber Defence Systems (ACDS). These systems, leveraging the power of artificial intelligence and machine learning, promise a revolution in the cybersecurity landscape, responding to threats and attacks with unprecedented speed and efficiency. But understanding this trend requires an exploration of the concept, potential benefits, challenges, and ethical implications.

Autonomous Cyber Defence Systems represent the pinnacle of AI application in cybersecurity. They are systems that can automatically detect, evaluate, and respond to cyber threats without human intervention. They envision a future where cybersecurity systems are not just responsive but proactive, anticipating threats before they occur and implementing countermeasures.

The potential benefits of such systems are significant. Firstly, the speed of response can be drastically improved. Manual threat identification and response can be slow and prone to errors. ACDS, on the other hand, can analyse vast amounts of data in real-time, identifying and neutralising threats at an impressive speed, often before any significant damage is done.

Secondly, ACDS can handle the increasing volume and complexity of cyber threats. The cybersecurity landscape is evolving rapidly, with threats becoming more sophisticated and frequent. Humans alone, regardless of their expertise, cannot effectively handle this influx. Autonomous systems can scale to meet this challenge, continually learning from new data to improve their threat detection and response capabilities.

However, the shift towards autonomy is not without challenges. One of the primary concerns is the risk of false positives and negatives. While AI has made significant strides, it is not fool proof. Misinterpretation of benign activities as threats, or overlooking actual threats, can have serious consequences.

Additionally, autonomous systems may become targets themselves. Cybercriminals are increasingly trying to exploit AI systems, either by feeding them misleading data to induce incorrect decisions (poisoning attacks) or by probing them to understand their inner workings and find weaknesses (evasion attacks).

An equally important aspect is the ethical and legal considerations. Who is responsible if an autonomous system makes a mistake that results in significant damage? How do we ensure that these systems respect privacy laws when they collect and analyse data?

There's also the potential risk of job displacement. As ACDS become more common, there will be a shift in the cybersecurity job landscape. However, rather than eliminating jobs, it's likely to redefine them. Cybersecurity professionals will need to adapt, developing skills to design, implement, and oversee these autonomous systems.

In conclusion, the future of ACDS is poised on the edge of significant challenges and opportunities. The potential for autonomous systems to transform cybersecurity is immense, but realising this potential will require careful navigation of technical, ethical, and legal hurdles. As we move towards this future, a balanced, thoughtful approach is key – one that leverages the strengths of ACDS while acknowledging and addressing their limitations.

11.3. Quantum Computing and Cybersecurity

Quantum computing is a revolutionary technology, brewing on the horizons of technological advancements. This phenomenon, a veritable juggernaut of computational power, harnesses the unique properties of quantum physics to perform complex calculations at speeds that are unfathomable with today's conventional computers. But this leap forward in technology is not without implications for cybersecurity – a double-edged sword that presents both challenges and opportunities.

As we explore this new frontier, we must tread lightly, for the terrain is both promising and perilous. Quantum computers could potentially crack encryption algorithms that current systems deem secure. Today's encryption methods, such as RSA and ECC, base their security on the complexity of factoring large numbers into primes - a feat easily achievable by quantum machines. With quantum computers, these systems could be cracked open in mere seconds, leaving our data and digital infrastructure vulnerable. This is not a distant threat, but a future reality that the cybersecurity world refers to as "Quantum Apocalypse."

Yet, in this complex tapestry of quantum computing, a thread of hope weaves its way through. The same power that could break current encryption can be harnessed to create quantum-resistant algorithms, a new breed of encryption that can withstand quantum attacks. This field, known as post-quantum cryptography, is still in its infancy but is rapidly gaining traction.

Moreover, quantum computing could provide extraordinary new tools for cybersecurity. Quantum key distribution (QKD) leverages the principles of quantum mechanics to enable secure communication. Any attempt to intercept the quantum key changes its state, revealing the presence of the intruder. Thus, quantum mechanics may provide a way to secure communications in a manner that is impossible for classical computers to violate.

This paradox of quantum computing is a testament to the intertwining of risk and opportunity in technological advancement. As we step forward into this quantum future, we must engage in a delicate dance - embracing the possibilities that quantum computing brings to cybersecurity, while also preparing for the challenges. It requires the cybersecurity field to engage with quantum computing proactively, fostering collaboration between quantum physicists, computer scientists, and cybersecurity experts. This collaboration is not a luxury but a necessity in the face of a technology that could redefine the very fabric of cybersecurity.

11.3.1. Understanding Quantum Computing: Potential and Perils

Quantum computing is a technological realm marked by transformative potential and inherent perils. This domain remains enigmatic to most, yet understanding it is vital to our future, as it poses both an unprecedented threat and a novel opportunity to cybersecurity.

At its essence, quantum computing is the exploitation of quantum mechanics to perform computational tasks. A quantum computer uses qubits instead of bits to represent and manipulate information. Unlike a bit, which can be either 0 or 1, a qubit can be both 0 and 1 at the same time, thanks to the quantum phenomena of superposition. Additionally, qubits can be entangled due to quantum entanglement, meaning the state of one qubit can instantaneously affect the state of another, regardless of the distance separating them. These unique properties imbue quantum computers with enormous computational power, enabling them to solve complex problems and perform calculations far beyond the capabilities of classical computers.

Yet, this immense power also constitutes the primary peril of quantum computing. Today's cybersecurity infrastructure relies heavily on encryption algorithms based on the computational difficulty of factoring large numbers or finding discrete logarithms - tasks that classical computers find time-consuming. But a sufficiently powerful quantum computer, using Shor's algorithm, could accomplish these tasks with relative ease, rendering our current cryptographic defences obsolete. The RSA encryption, elliptic-curve cryptography (ECC), and digital signatures would all succumb to the might of quantum computers. This vulnerability is termed the "Quantum Threat," and it presents a substantial challenge for cybersecurity.

Conversely, quantum computing also offers novel ways to counteract these threats. The quantum world's intrinsic properties give rise to Quantum Key Distribution (QKD), an ultra-secure communication method. QKD enables two parties to generate a secret key that can be used for encrypted communication, and any eavesdropping attempt will disturb the quantum system and reveal the intruder. Furthermore, efforts are underway to devise quantum-resistant algorithms that can withstand attacks from quantum computers, paving the way for post-quantum cryptography.

Understanding quantum computing is like navigating a two-faced labyrinth. One path leads to unprecedented cybersecurity threats, potentially undermining our current encryption systems. The other path guides us toward novel cybersecurity solutions, exploiting quantum principles to devise ultra-secure communication protocols and resilient encryption algorithms. The challenge lies in leveraging the potential while mitigating the perils - a task that requires an interdisciplinary approach, coupling the expertise of cybersecurity professionals, quantum physicists, and computer scientists.

11.3.2. Quantum Computing and Cryptography: The Coming Quantum Apocalypse?

As the conversation around quantum computing intensifies, so does the fear of a potential 'Quantum Apocalypse'—a future where our current cryptographic security systems become obsolete in the face of the sheer computational power of quantum computers.

Indeed, the bedrock of present-day cryptography—public-key cryptosystems like RSA and elliptic-curve cryptography (ECC)—is under threat. These systems derive their security from the difficulty of factoring large numbers or solving discrete logarithm problems, tasks which, for classical computers, would require an inordinate amount of time to complete. However, quantum computers, using algorithms such as Shor's algorithm, could potentially solve these problems significantly faster, effectively breaking the security of these cryptosystems.

The ramification of such a development is immense. The majority of the online communication today relies on these cryptographic systems to ensure confidentiality, integrity, and authenticity of data. A powerful, large-scale quantum computer could decrypt sensitive information, from personal data to state secrets, exposing a vast expanse of confidential information and creating unprecedented chaos, hence the term 'Quantum Apocalypse'.

However, before we concede defeat, it's crucial to remember that this potential 'apocalypse' is not imminent. Building a large-scale, fault-tolerant quantum computer is an extraordinary technological challenge and, according to most experts, we are still a few decades away from such a development. This provides us with a window of opportunity to prepare and fortify our defences.

In response to this looming threat, the field of post-quantum cryptography has emerged, focusing on designing new cryptographic algorithms that could withstand quantum attacks. These algorithms are based on mathematical problems that are believed to be resistant to quantum computing techniques. For example, lattice-based cryptography, code-based cryptography, and multivariate cryptography are some promising directions in this area.

Furthermore, quantum computing itself offers new avenues for securing data. Quantum Key Distribution (QKD), for instance, capitalises on the principles of quantum mechanics to exchange cryptographic keys, ensuring secure communication. Any attempt at eavesdropping on the key exchange process will inherently disturb the quantum states involved and alert the communicating parties of the breach.

In conclusion, while quantum computing poses a real threat to current cryptographic systems, the 'Quantum Apocalypse' is not a foregone conclusion. It represents a clear call to action for researchers, technologists, and policymakers to invest time and resources in post-quantum cryptography and quantum-based security solutions. It underscores the need for a proactive approach in reshaping our cybersecurity infrastructure, an endeavour that requires global cooperation, diligent research, and innovative thinking.

11.3.3. Preparing for the Quantum Future: Post-Quantum Cryptography

In the light of quantum computing's impending challenge to classical cryptography, one might wonder: how do we prepare for such an event? The answer lies in a burgeoning field called Post-Quantum Cryptography (PQC), also known as Quantum-Resistant Cryptography. Its goal? To develop cryptographic systems that can withstand the computational power of quantum computers, effectively future proofing our digital security architecture.

Post-Quantum Cryptography draws upon complex mathematical problems that, unlike factoring or the discrete logarithm problem, are believed to remain hard even for quantum computers. Many promising cryptographic techniques underpin this field, and it is upon these we shall turn our attention.

One of the significant contenders in the PQC realm is lattice-based cryptography. Lattices, which are multi-dimensional grids of points, pose mathematical problems considered difficult for both classical and quantum computers. A lattice's inherent structure, particularly in high dimensions, proves challenging to navigate computationally, even with quantum algorithms.

Another promising candidate in the PQC arena is code-based cryptography, such as the McEliece cryptosystem. It relies on the hardness of decoding a general linear code, a problem which is NP-hard and not known to be solvable by quantum computers. However, a notable drawback of this system is the size of the keys, which are substantially larger than those used in conventional cryptography.

Hash-based cryptography forms yet another part of the PQC portfolio. This form of cryptography leverages cryptographic hash functions—widely used in classical cryptography—to create secure signatures. They are currently the most mature and well-understood PQC algorithms, but the drawback lies in their limitation to signature schemes, unable to offer key exchange or encryption capabilities.

Also in the post quantum mix are multivariate cryptography and isogeny-based cryptography, each with their unique approaches to resisting quantum attacks. The former involves solving systems of multivariate polynomials, while the latter exploits the computational difficulty of certain problems in the arithmetic of elliptic curves.

While these techniques show promise, transitioning to post-quantum cryptography is a significant undertaking. New cryptographic systems must undergo rigorous vetting for potential vulnerabilities, and widespread deployment would require substantial changes to existing digital infrastructures. Furthermore, there will be trade-offs to consider increased key sizes, changes in encryption speeds, and compatibility issues, to name a few.

It's also crucial to act pre-emptively. Quantum computers capable of breaking current cryptographic systems may still be some years away, but information encrypted today could be intercepted and stored for decryption by future quantum computers—a prospect known as the "harvest now, decrypt later" attack.

11.4. The Future of Cybersecurity Careers and Skills

To stay ahead of potential threats, they must be agile, lifelong learners, ready to continuously adapt and grow.

The growing sophistication of cyber threats has led to a demand for more specialised roles in the industry. Cybersecurity careers are not limited to the generic role of 'security analyst' or 'security engineer'. The landscape has broadened, bringing forth niche roles like cloud security specialists, industrial control system security analysts, and post-quantum cryptographers. These roles require an in-depth understanding of specific technologies or sectors, marking a move from generalists to specialists.

Moreover, artificial intelligence and machine learning are transforming cybersecurity practices, opening up new roles like AI security specialists. These professionals must not only understand how to use AI to detect and mitigate threats but also how to protect AI systems themselves, a new field dubbed "AI security."

On the flip side, with the advent of quantum computing, there will be a demand for professionals skilled in post-quantum cryptography. These specialists will be at the forefront of developing and implementing cryptographic solutions that can withstand the power of quantum computers.

Notably, the technical skills required for cybersecurity careers are shifting as well. While a solid foundation in IT and computing remains critical, skills like data analysis and understanding of machine learning algorithms are becoming increasingly valuable. Professionals need to understand not just how to defend systems but how potential attackers might leverage new technologies for nefarious purposes.

Furthermore, cybersecurity professionals are required to have an increasingly robust understanding of regulatory and compliance issues. As data privacy laws evolve worldwide, professionals need to ensure that cybersecurity practices align with legal and ethical requirements. This confluence of law, ethics, and cybersecurity creates a need for professionals who can navigate these intersecting domains.

On the softer side, the value of communication skills in cybersecurity roles cannot be overstated. Cybersecurity professionals often have to explain complex technical issues to non-technical stakeholders. They need to articulate risks, strategies, and the potential impact of cybersecurity issues clearly and persuasively.

Also, the ever-present cybersecurity skills gap is likely to persist into the future. The increasing demand for skilled cybersecurity professionals outstrips supply, leading to numerous unfilled roles. This gap presents a significant opportunity for those considering a career in this field, promising job security and competitive salaries.

In essence, the future of cybersecurity careers will be characterised by specialisation, a blend of evolving technical and soft skills, and a strong demand for professionals. As the industry grapples with emerging threats, the need for a highly skilled and versatile cybersecurity workforce will remain a constant.

11.4.1. Emerging Cybersecurity Roles and Skill Sets

The exponential growth of connected devices due to the Internet of Things (IoT) has necessitated the role of IoT Security Specialists. These professionals are tasked with securing an ever-growing network of devices. They need to be well-versed in wireless communications, real-time operating systems, and IoT security protocols.

Another emerging role is that of the Privacy Engineer. As organisations wrestle with stringent data protection regulations, the role of Privacy Engineers is becoming more critical. They are entrusted with the job of embedding privacy considerations into product design and digital services, adhering to the principle of Privacy by Design. Strong knowledge of global data protection laws, familiarity with data management, and understanding of security architectures are essential for this role.

As artificial intelligence and machine learning become mainstream in cybersecurity, we see the emergence of AI Security Specialists. They are expected to secure AI systems from adversarial attacks and use AI to enhance security defences. Professionals in this role require strong knowledge of AI/ML principles and algorithms, and a thorough understanding of the cybersecurity landscape.

On the horizon, we see roles such as Quantum Cryptographers. These professionals will be crucial in developing encryption that can withstand attacks from quantum computers. To work in this role, one needs an in-depth understanding of quantum computing principles and cryptography.

The skill sets required for these emerging roles are as varied as the roles themselves. Soft skills, though, remain universally relevant. For instance, critical thinking is indispensable in the cybersecurity profession. Whether it's identifying vulnerabilities, detecting anomalous patterns, or devising defensive strategies, the ability to think critically is crucial.

Problem-solving skills are another necessity. Cybersecurity professionals often find themselves in situations where they need to find a quick and effective solution to a security problem. Being able to navigate such scenarios is key.

Finally, communication skills are becoming increasingly important. As cybersecurity becomes a boardroom issue, professionals need to be able to translate complex cybersecurity jargon into business language. They should be able to clearly communicate risks, strategies, and mitigation plans to stakeholders without overwhelming them with technical details.

In conclusion, the future of cybersecurity careers appears vibrant and multifaceted. There's a growing demand for varied skills and roles in the cybersecurity field, making it an attractive career choice for many. As the landscape of cyber threats continues to evolve, so too will the roles and skills needed to counter them, providing numerous opportunities for those ready to rise to the challenge.

11.4.2. The Role of Professional Certifications in Cybersecurity Careers

Professional certifications have multiple roles within cybersecurity careers. They offer an objective validation of a professional's knowledge and skills, giving them an edge in the job market. For employers, certifications help identify candidates who possess specific skill sets needed to protect the organisation's digital assets effectively.

For instance, the Certified Information Systems Security Professional (CISSP) certification, offered by the International Information System Security Certification Consortium (ISC)², is considered a gold standard within the industry. Holding a CISSP certification signifies that a professional possesses advanced skills in designing, implementing, and managing a cybersecurity program.

Similarly, the Certified Ethical Hacker (CEH) certification, offered by the EC-Council, validates a professional's understanding of how to think and operate like a hacker, providing the skills needed to lawfully test and bolster an organisation's security systems.

As cybersecurity threats continue to evolve, new certification programs are emerging in response. For instance, certifications focused on cloud security, such as the Certified Cloud Security Professional (CCSP) certification, are gaining prominence as businesses increasingly move their operations to the cloud.

Additionally, certifications like the Certified Information Privacy Professional (CIPP), which focuses on data privacy, are becoming valuable as businesses navigate complex privacy laws and regulations like the GDPR and CCPA.

However, while professional certifications are beneficial, they are not a panacea. They should be viewed as part of a broader professional development strategy, complementing on-the-job experience and ongoing education. No certification can replace the insights gained from real-world experience, the understanding developed from solving practical problems, or the adaptability derived from navigating the ever-changing cybersecurity landscape.

It's also worth noting that the journey to attaining these certifications often involves rigorous study and examination, which can enhance a professional's depth of knowledge. Preparing for a certification exam can provide a structured learning path, covering various domains of cybersecurity, and ensuring a comprehensive understanding of the field.

In conclusion, professional certifications play a pivotal role in cybersecurity careers. They serve as important milestones in a professional's journey, showcasing their commitment to the field, and their willingness to invest in their skills and knowledge. As we move into the future, these certifications will continue to evolve, mirroring the changing landscape of cybersecurity threats and the new skills required to mitigate them.

11.4.3. Fostering a Culture of Cybersecurity in Organisations

As we venture deeper into an era defined by digitalisation, it is increasingly evident that robust cybersecurity is not simply a function of having the right technology or even the best-skilled professionals in place. Instead, a robust cybersecurity posture is, in essence, a cultural factor. It rests on the collective actions and attitudes of every individual within an organisation, from the C-suite to the newest recruit.

This awareness forms the bedrock of a culture of cybersecurity. Such a culture manifests as a set of shared attitudes, values, goals, and practices that characterise an organisation's approach to its information and systems' integrity, confidentiality, and availability. It's an ecosystem where cybersecurity becomes a habitual consideration, ingrained into the fabric of daily operations and strategic planning.

One key step towards fostering a cybersecurity culture is making cybersecurity an executive issue. Leadership plays a pivotal role in setting the tone for cybersecurity. When executives demonstrate a commitment to cybersecurity, it sends a clear message throughout the organisation that cybersecurity is not an afterthought but a core business function. Leadership should be involved in defining cybersecurity strategies, understanding the threat landscape, and making informed decisions about risk tolerance and resource allocation.

Education is another crucial pillar in building a cybersecurity culture. Continuous training programs that teach employees about potential threats, safe online behaviour, and response plans in case of a cybersecurity incident are essential. However, it's important that these training sessions are not one-time events but continuous efforts, reflecting the dynamic nature of the cybersecurity landscape. Cybersecurity drills, just like fire drills, can be conducted to ensure that employees understand their roles and responsibilities during a cybersecurity incident.

Encouraging openness and communication is also integral to a cybersecurity culture. Employees should feel comfortable reporting potential issues, such as a suspicious email, without the fear of repercussions. By encouraging a proactive stance, organisations can often nip threats in the bud before they escalate.

Finally, a culture of cybersecurity recognises that mistakes will happen and uses them as learning opportunities. Instead of fostering a fear of reprisal, organisations should foster a sense of shared responsibility for security. When cybersecurity is seen not as a burdensome obligation but as a shared duty, everyone becomes a part of the defence.

As we look towards the future of cybersecurity, the culture within organisations will become even more significant. As threats become more sophisticated and the attack surface continues to expand, a well-informed, security-conscious workforce could well be the most potent defence against cyber adversaries. In this regard, fostering a culture of cybersecurity is not just a strategic move, but an essential one.

CHAPTER 12. CONCLUSION: CYBERSECURITY AS A LIFELONG PURSUIT

This chapter serves as a synthesis of the themes and insights we have explored throughout this book, threading together the various elements of cybersecurity to present a comprehensive overview of the field. It underscores the need for continuous learning, constant vigilance, and an ever-evolving understanding of threats and best practices. This is not a domain where one can ever afford to rest on their laurels or become complacent with past knowledge.

It's important to remember that the cyber landscape is perpetually changing. New technologies introduce new vulnerabilities; novel threats emerge as old ones are neutralised. Staying ahead of these changes necessitates a commitment to lifelong learning. It requires one to keep pace with emerging trends, technologies, and tactics, and continuously upskill to meet the evolving demands of the cyber realm.

As we wrap up this comprehensive exploration of cybersecurity, we hope that readers will walk away with an understanding that cybersecurity is more than a set of technical skills or a regulatory compliance requirement. Rather, it is a mindset, a commitment, and an ongoing pursuit. It's a journey, not a destination. The conclusion of this book is but the beginning of your unique path in the dynamic world of cybersecurity.

12.1. Importance of Continuous Learning in Cybersecurity

What was cutting-edge understanding yesterday may well be outdated today. The breath-taking speed at which technology evolves, coupled with the incessant innovation in malicious tactics and attack methodologies, renders static knowledge both obsolete and perilous. Cybersecurity is a race without a finish line, where participants must continually adapt and evolve to avoid falling behind. Therefore, embracing the importance of continuous learning in cybersecurity isn't just advisable, it's absolutely essential.

Continuous learning serves multiple purposes in cybersecurity. It keeps professionals apprised of the latest threats, exposes them to the newest defence strategies, and ensures that their skills remain relevant in a rapidly evolving field. Furthermore, this consistent cycle of learning and application engenders a proactive, rather than reactive, approach to cybersecurity. It empowers individuals and organisations to anticipate potential threats and implement protective measures in advance, thereby minimising the chances of a breach and limiting the damage when one does occur.

Moreover, continuous learning is key to understanding the nuances of new technologies and applications. Consider the advent of cloud computing, AI, and IoT devices; each of these developments has introduced fresh vectors of vulnerability. Professionals equipped with up-to-date knowledge about these technologies are better prepared to secure them against potential threats. As such, continuous learning ensures that cybersecurity keeps pace with technological innovation.

Beyond the practical benefits, continuous learning in cybersecurity also serves to foster a culture of curiosity, resilience, and innovation. It is through this perpetual quest for knowledge that new ideas emerge, ground-breaking solutions are conceived, and existing systems are continually refined and improved. In this context, continuous learning becomes an act of empowerment, enabling each participant to contribute meaningfully to the broader discourse on cybersecurity.

Finally, continuous learning also enhances the career prospects of cybersecurity professionals. As employers seek experts who can navigate the complexities of today's threat landscape, those who demonstrate an ongoing commitment to learning are more likely to secure advanced roles and responsibilities. By continually expanding their knowledge and skills, professionals can not only stay relevant in this dynamic field, but they can also differentiate themselves in a competitive job market.

In summary, continuous learning is an indispensable facet of the cybersecurity domain. It is the driving force that ensures progress, underpins resilience, and facilitates innovation, making it an essential practice for anyone invested in the field of cybersecurity.

12.1.1. The Ever-evolving Nature of Cyber Threats

Cyber threats evolve for many reasons. For one, the vast and global nature of the digital world offers a fertile ground for criminal activities. With an increasingly interconnected world, the potential points of vulnerability multiply, offering numerous opportunities for exploitation. In this complex environment, every new technology, application, or service can open doors to novel forms of cyber threats

Moreover, threat actors have become more sophisticated, resourceful, and organised. They now use advanced techniques like artificial intelligence and machine learning to conduct more targeted and impactful attacks.

Cybercriminals are also becoming more adept at leveraging social engineering tactics, exploiting the human factor, which is often the weakest link in cybersecurity.

Additionally, the motives behind cyber threats are becoming more varied. While financial gain remains a significant driver, cyberattacks now also aim to cause disruption, sway public opinion, or steal sensitive information for strategic advantage. This broadening of objectives has led to the emergence of state-sponsored cyber threats, making the cyber landscape an extension of geopolitical conflicts.

Lastly, the ever-evolving nature of cyber threats is also fuelled by an arms race between cybersecurity professionals and threat actors. Every time a new defensive measure is developed, attackers work tirelessly to find a way around it. As long as there is value in digital assets, there will always be those seeking unauthorised access to them.

In this challenging environment, an appreciation for the ever-evolving nature of cyber threats is crucial. It serves as a stark reminder that cybersecurity is not a one-time effort but an ongoing commitment. It reinforces the importance of staying abreast of the latest trends, continually updating and refining security measures, and adopting proactive and dynamic approaches to cybersecurity. After all, in a race against adversaries that never stand still, standing still is not an option.

12.1.2. Lifelong Learning Strategies for Cybersecurity Professionals

One of the fundamental lifelong learning strategies is maintaining a strong commitment to professional development. This includes engaging in various formal educational opportunities. Pursuing advanced degrees, attending workshops and seminars, and obtaining professional certifications such as CompTIA Security+, Certified Information Systems Security

Professional (CISSP), and Certified Ethical Hacker (CEH), are vital. These programs keep professionals apprised of the latest technologies, strategies, and regulatory changes in cybersecurity.

Self-study is another crucial element of lifelong learning. With the proliferation of online resources such as Massive Open Online Courses (MOOCs), webinars, podcasts, and eBooks, cybersecurity professionals can gain a wealth of knowledge at their convenience. Keeping an eye on research papers and technical reports from prominent cybersecurity organisations and research institutions can also provide valuable insights into emerging threats and innovative solutions.

Participating in hands-on experiences, such as Capture The Flag (CTF) competitions, hackathons, or real-life simulations, offers practical insights and hones problem-solving skills. This approach helps professionals apply theoretical concepts to real-world situations, enhancing their understanding and proficiency in handling various cybersecurity scenarios.

Networking with other professionals in the field is another learning strategy. Joining cybersecurity associations, attending conferences, or participating in online forums allows professionals to exchange ideas, learn from each other's experiences, and stay aware of emerging trends and challenges. This collaborative approach broadens their perspective, making them better prepared to tackle new threats.

Developing a habit of reflective practice is also crucial. By routinely assessing their work, cybersecurity professionals can identify gaps in their knowledge, pinpoint areas for improvement, and set learning goals. This practice can guide their learning journey and help them become more effective in their roles.

In addition, considering ethics and a broader societal impact of cybersecurity measures is essential. Cybersecurity professionals must understand the ethical implications of their actions and consider the effects of cyber technologies on society. This ethical dimension of learning helps ensure their actions align with legal requirements and societal expectations.

Finally, embracing curiosity and a passion for innovation can fuel the learning journey. Cybersecurity, at its core, is about finding novel solutions to complex problems. Therefore, nurturing an innovative mindset can keep professionals ahead of the curve and enable them to devise more effective strategies to counter evolving threats.

12.1.3. Role of Certifications and Continuous Education in Cybersecurity

Cybersecurity, with its fast-paced evolution, presents a continuous learning challenge for professionals in the field. Two instrumental aspects supporting this quest for knowledge are certifications and continuous education. Both offer a pathway to stay current with new technologies, understand evolving threats, and remain informed about industry standards.

Certifications hold a vital place in the cybersecurity landscape. They validate the skills and knowledge of professionals, providing a benchmark for proficiency in various cybersecurity domains. Certifications such as CompTIA Security+, Certified Information Systems Security Professional (CISSP), Certified Ethical Hacker (CEH), and Certified Information Privacy Professional (CIPP), amongst others, encompass a wide array of focus areas from network security to privacy law and regulations.

Each certification has a different emphasis, target audience, and prerequisites. For example, CISSP is designed for seasoned professionals aiming for leadership roles, while Security+ may be more suitable for those starting their cybersecurity careers. Holding these certifications not only enhances professional credibility but also ensures a continuous learning process. This is because they usually require individuals to earn continuing professional education (CPE) credits to maintain their validity.

Continuous education is another cornerstone of lifelong learning in cybersecurity. It allows professionals to keep abreast of technological advancements, regulatory changes, and evolving cyber threats. This education can take many forms – from formal degree programs offered by universities to online courses, seminars, workshops, and training sessions conducted by reputed cybersecurity organisations.

Many academic institutions now offer graduate programs in cybersecurity, allowing professionals to delve deeper into specific areas of interest. These courses often combine theoretical knowledge with practical exercises, thus providing a well-rounded learning experience. On the other hand, online platforms offer flexible learning opportunities that professionals can access at their convenience, often without having to interrupt their working life.

Additionally, continuous education programs are essential for staying aware of regulatory changes in the field. As privacy laws and data security regulations evolve worldwide, it is imperative for cybersecurity professionals to understand these changes and their implications on business practices.

In the broader perspective, the symbiotic relationship between certifications and continuous education creates an environment conducive to lifelong learning. Certifications guide the learning process by setting industry-recognised standards and benchmarks, while continuous education provides the resources and opportunities to reach these benchmarks and beyond.

The essence of cybersecurity lies not only in defending against threats but also in fostering a commitment to lifelong learning. As cybersecurity professionals navigate the intricacies of this challenging field, certifications and continuous education will continue to be their invaluable companions.

12.2. Staying Updated on the Latest Cybersecurity Trends

Emerging cybersecurity trends encompass a broad range of areas including new technologies, attack methodologies, legislative changes, and evolving threat landscapes. Understanding these trends allows professionals to anticipate potential risks, devise effective defence strategies, and ensure that their cybersecurity skill set remains relevant in a rapidly changing environment.

Technological advancements frequently reshape the cybersecurity landscape. From the advent of cloud computing to the rise of Internet of Things (IoT) devices, each new wave of technology brings with it unique cybersecurity challenges. For instance, the accelerating adoption of quantum computing presents both remarkable potential for enhanced security measures, and potential perils in terms of its ability to disrupt current encryption standards. By staying updated on such technology trends, cybersecurity professionals can better understand these emerging threats and devise strategies to mitigate them.

Simultaneously, the ways in which cyberattacks are conducted are also continually evolving. Advanced Persistent Threats (APTs), ransomware attacks, and social engineering techniques are becoming increasingly sophisticated, often outpacing traditional defence mechanisms. Awareness of such trends is key to developing proactive, rather than merely reactive, defence strategies.

Furthermore, as the significance of data protection grows, so does the complexity and scope of related legislation. Data protection laws such as the European Union's General Data Protection Regulation (GDPR) and the California Consumer Privacy Act (CCPA) have profound implications for how organisations handle data security. Keeping abreast of these changes is vital to ensure legal compliance and to protect an organisation's reputation.

The cybersecurity industry offers various resources to stay updated on these trends. These include industry reports, webinars, online forums, and professional networks. Additionally, attending cybersecurity conferences, subscribing to industry journals, and following thought leaders on social media can also provide valuable insights.

12.2.1. Following Trusted Sources for Cybersecurity News and Research

Navigating the sea of information in cybersecurity can be a daunting task, given the speed at which developments occur. Following trusted sources for news and research, however, can significantly streamline this process, ensuring you remain at the cutting edge of your field.

In the cybersecurity world, industry-focused news sites, academic journals, and blogs written by respected professionals are valuable resources. These offer a wealth of information, providing comprehensive coverage of the most recent advancements, evolving threats, and regulatory changes. Websites like Krebs on Security, DarkReading, or The Hacker News deliver up-to-the-minute news on security incidents and threats, while Schneier on Security and the Google Security Blog offer deeper insights into cybersecurity trends and technologies.

Academic journals such as the Journal of Cybersecurity, Computers & Security, and the Journal of Information Security offer rigorously peer-reviewed research, presenting some of the most advanced work in the field. These sources are especially useful for those wanting to delve deeper into the theoretical and technical aspects of cybersecurity.

In addition to these, threat intelligence feeds like those provided by CrowdStrike or Recorded Future can give a near real-time perspective on active threats. These services aggregate data from a multitude of sources and use advanced analytics to identify emerging threats, providing actionable intelligence that can inform an organisation's defensive strategies.

Online forums and communities such as StackExchange's Information Security, Reddit's r/netsec, or GitHub also present rich platforms for knowledge sharing and discussion. These spaces allow professionals to exchange ideas, discuss recent developments, and even crowdsource solutions to complex problems.

Lastly, following the security advisories and blogs of software and hardware vendors is crucial. Companies like Microsoft, Cisco, and IBM routinely publish advisories on vulnerabilities in their products, along with recommended mitigations and patches.

It is not enough, however, to merely follow these sources. Critical evaluation of the information is paramount, given the potentially severe implications of acting on incorrect or misleading information. A keen understanding of the source's reputation, expertise, and potential biases helps in assessing the validity and relevance of the information presented.

Following trusted sources for cybersecurity news and research, therefore, is a multifaceted process. It involves not just identifying and keeping track of these sources, but also critically analysing the information they provide. Doing so helps cybersecurity professionals stay well-informed, well-prepared, and well-equipped to handle the evolving threats that their organisations face.

12.2.2. Engaging with the Cybersecurity Community: Conferences, Forums, and Social Media

Active engagement with the global cybersecurity community forms a pivotal part of a cybersecurity professional's journey. This proactive involvement broadens one's horizons, facilitates networking, and propels individuals to stay updated on recent trends. It is through a variety of channels like conferences, forums, and social media that this engagement often unfolds.

Cybersecurity conferences represent a melting pot of ideas, discoveries, and collaborative opportunities. They bring together professionals, academics, researchers, and enthusiasts from all corners of the globe, providing a platform for knowledge exchange on a massive scale. Events such as the RSA Conference, Black Hat, DEFCON, and the Cybersecurity & Cloud Expo offer a mix of keynote speeches, technical presentations, workshops, and panel discussions, each shedding light on different facets of cybersecurity. These events often serve as launch pads for ground-breaking research and innovative technologies, making them invaluable for those keen on keeping pace with the industry.

Forums and discussion boards present another avenue for dynamic engagement with the cybersecurity community. Spaces like StackExchange, Reddit, and GitHub offer an environment where experts and novices alike can pose questions, share insights, and collaborate on projects or problem-solving. These platforms, due to their accessibility and user-driven content, can be remarkably effective in promoting continuous learning and growth.

The role of social media in engaging with the cybersecurity community cannot be overstated. Platforms like Twitter and LinkedIn serve as bustling digital hubs where professionals can connect, share content, and partake in discussions. Following influential figures, participating in relevant groups, and joining industry-specific hashtags can keep one attuned to the pulse of the cybersecurity sphere.

Podcasts and webinars, like those offered by the CyberWire or Darknet Diaries, also create engaging, accessible content, presenting an excellent resource for keeping up to date with industry trends, learning from experts, and exploring various cybersecurity topics in depth.

In addition to knowledge sharing and networking, these engagements often provide emotional support, encouraging a sense of camaraderie in the face of shared challenges. They remind us that, although the task of securing our digital world can seem daunting, we are not alone in our endeavours.

However, as we engage with this vast community, it is critical to maintain a level of discernment, separating the wheat from the chaff. Not all information or advice will be beneficial or accurate. Verifying information, cross-checking sources, and maintaining a healthy scepticism are key practices to adopt in this journey of continuous learning and engagement.

In essence, engaging with the cybersecurity community across conferences, forums, and social media serves multiple purposes - knowledge expansion, networking, collaboration, emotional support, and staying abreast of the rapid changes in the field. It is a cornerstone in a cybersecurity professional's continuous learning journey.

12.3. Cybersecurity as a Shared Responsibility

From individual internet users to global tech giants, from the smallest start-ups to massive government agencies, everyone has a role to play in maintaining digital security and safeguarding sensitive data.

At an individual level, understanding and practicing basic cybersecurity hygiene such as strong, unique passwords, two-factor authentication, regular software updates, and awareness of phishing and other online scams, can provide a robust first line of defence against cyber threats. An educated user can significantly reduce the risks of data breaches and identity theft.

Within organisations, cybersecurity is not merely the concern of IT or security teams; rather, it extends to every employee. Whether it's the CEO, the HR manager, or the office intern, each person can potentially be a weak link or a strong safeguard, depending on their cybersecurity awareness and practices. A culture of security can effectively mitigate the risk of insider threats, which often arise not from malice but ignorance or carelessness.

At an industry level, collaboration is critical in responding to shared threats. Industries often face similar cybersecurity challenges, and cooperation, sharing of threat intelligence and best practices can significantly strengthen the collective security posture. ISACs (Information Sharing and Analysis Centres) are an excellent example of industry-specific organisations that promote such cooperation.

At a national and international level, governments have a crucial role in defining laws and regulations for cybersecurity, facilitating information sharing, and coordinating responses to large-scale cyber threats. Collaboration between nations can help to deter state-sponsored cyber threats and bring cybercriminals to justice.

Technology developers and service providers also carry the responsibility of creating secure products and services, prioritising security in every stage of design and deployment, and responding quickly to identified vulnerabilities. Moreover, they must respect user privacy, upholding ethical considerations alongside functional ones.

12.3.1. Everyone's Role in Ensuring Cybersecurity

In the vast and complex universe of digital interactions, the assurance of cybersecurity is not a one-person show, nor a task confined to the perimeters of IT departments. Instead, cybersecurity emerges as an inclusive symphony, requiring the coordinated effort of every participant, every role, and every stakeholder within this expansive digital network.

Every individual interacting with the digital space holds a piece of the cybersecurity puzzle. The casual internet user, for instance, shapes the frontline of cybersecurity defence. Their alertness to phishing attempts, respect for strong password protocols, willingness to perform regular software updates, and general awareness of cybersecurity best practices form the first barricade against a multitude of cyber threats. This individual vigilance can deter common cybersecurity threats, protect personal data, and even shield entire networks from potential breaches.

The baton of cybersecurity is also firmly held within organisational structures, reaching far beyond the confines of dedicated IT teams. Every employee within an organisation can either strengthen the protective perimeter or unknowingly create a breach. Leaders and managers have a responsibility to foster a proactive culture of cybersecurity, making it as commonplace as any other organisational protocol. Routine cybersecurity training, open communication about potential threats, and mutual accountability can turn each team member into a human firewall.

On a broader canvas, industry-wide cooperation significantly bolsters collective defences. When competitors unite against shared cybersecurity threats, they strengthen their individual as well as mutual resilience. By creating platforms for knowledge exchange, like Information Sharing and Analysis Centres (ISACs), industries can pool resources, share threat intelligence, and learn from each other's experiences to construct a unified front against cyber threats.

Zooming out further, we observe the crucial role played by national and international governments. These entities set the stage for cybersecurity norms, laws, and regulations, influencing the global cybersecurity landscape. By facilitating knowledge sharing, coordinating responses to large-scale threats, and setting legal consequences for breaches, they encourage responsible behaviour within the digital ecosystem.

Lastly, but most certainly not least, technology creators bear the significant responsibility of designing and deploying secure products and services. They must prioritise security at every stage of the development cycle, swiftly address vulnerabilities, and balance functionality with ethical considerations, particularly regarding user privacy.

In essence, cybersecurity is a shared script, where everyone must play their part diligently. The collective effort from all these individual roles can help construct a more secure and trustworthy digital theatre.

12.3.2. Cybersecurity Best Practices for Everyone: A Recap

Recapping cybersecurity best practices is not merely an exercise of revisiting previously discussed points but is an opportunity to emphasise that these practices form the bedrock of our collective resilience against cyber threats. Despite technological advancements and the complexity of the digital space, these fundamental principles remain pertinent to everyone navigating the cyber world.

First and foremost, we must pay heed to the adage, "knowledge is power." Continuous learning and staying informed about the latest threats, trends, and cybersecurity practices is vital. As the digital landscape evolves, so do the threat vectors and attack strategies. Regular training, webinars, cybersecurity news platforms, and online forums can be valuable resources for staying abreast of developments in the field.

Password hygiene is a fundamental practice in the digital world. Everyone should use strong, unique passwords across different accounts, updating them regularly. Utilisation of password managers and multi-factor authentication adds another layer of security, fortifying the access control to personal and organisational data.

Next, responsible internet behaviour includes vigilance against phishing and social engineering attacks. This involves a healthy scepticism towards unsolicited communication, careful inspection of emails and messages for signs of phishing and avoiding clicking on unverified links or attachments.

Regular software updates should be a standard practice. Developers continually improve and patch vulnerabilities in their software, hence keeping systems updated ensures the latest security defences are in place.

The principle of least privilege, which advises using the minimum necessary access rights for tasks, minimises the potential impact of a breach. This principle applies to all, from the individual user to the IT administrator.

Finally, backup strategies cannot be emphasised enough. Regularly backing up critical data ensures that, even in case of a ransomware attack or data corruption, recovery is possible.

Everyone has a part to play in the grand theatre of cybersecurity. Whether you're an individual internet user, an employee of an organisation, or a member of the cybersecurity community, adhering to these practices empowers not just your own cyber resilience, but fortifies the collective security of our interconnected digital ecosystem.

12.3.3. The Way Forward: Towards a More Secure Cyber World

As we step into the future, the cyber world continues to evolve at an unprecedented pace. This rapid evolution is a double-edged sword; while technological advancements bring immense benefits, they also expose us to an ever-increasing array of cyber threats. Hence, the journey towards a more secure cyber world is not a destination, but a constant pursuit that requires vigilance, agility, and a proactive mindset from all stakeholders.

The future of cybersecurity lies in the hands of those who understand the significance of their role in this cyber ecosystem. This role extends beyond just IT professionals and security experts. Everyone who interacts with the digital world, from individual internet users to corporate employees, needs to comprehend the impact of their actions on the overall security posture.

In this quest, education plays a pivotal role. Incorporating cybersecurity awareness into educational curriculums, professional training programs, and public awareness campaigns can empower users to safeguard their digital presence effectively. Additionally, initiatives promoting digital literacy can help bridge the gap between technological advancements and the average user's understanding, thereby reducing opportunities for cyber threats to exploit this disparity.

The intersection of cybersecurity with emerging technologies like AI, machine learning, and quantum computing presents a dynamic frontier. Leveraging these technologies can strengthen our defences and enable faster, more accurate threat detection and mitigation. However, as these technologies advance, so too do the capabilities of threat actors. Staying ahead of this curve necessitates continuous innovation, research, and collaboration within the cybersecurity community.

On a global scale, promoting cooperation between nations is crucial to combat cyber threats effectively. Cybersecurity challenges do not respect geographical boundaries. International treaties, agreements on cyber norms, information sharing platforms, and joint responses to cyber threats can significantly enhance our collective cybersecurity.

Furthermore, fostering a culture that values privacy and security is integral to building a more secure cyber world. Privacy must be considered not as an afterthought but as a prerequisite in digital activities and technological development. Implementing privacy-by-design and advocating for strong data protection regulations can help preserve individual privacy rights while still enabling digital innovation.

In conclusion, the path towards a more secure cyber world is paved with constant learning, proactive actions, collaboration, and an unwavering commitment to privacy and security. In this collective endeavour, every step we take in the right direction matters. By adopting a cybersecurity-first mindset, we can look forward to a future where we can harness the potential of the digital world confidently and securely.

www.ingramcontent.com/pod-product-compliance
Lightning Source LLC
Chambersburg PA
CBHW072026230526
45466CB00020B/932